Metal–Air and Metal–Sulfur Batteries

Fundamentals and Applications

ELECTROCHEMICAL ENERGY STORAGE AND CONVERSION

Series Editor: Jiujun Zhang
National Research Council Institute for Fuel Cell Innovation
Vancouver, British Columbia, Canada

Published Titles

Electrochemical Supercapacitors for Energy Storage and Delivery: Fundamentals and Applications
Aiping Yu, Victor Chabot, and Jiujun Zhang

Proton Exchange Membrane Fuel Cells
Zhigang Qi

Graphene: Energy Storage and Conversion Applications
Zhaoping Liu and Xufeng Zhou

Electrochemical Polymer Electrolyte Membranes
Jianhua Fang, Jinli Qiao, David P. Wilkinson, and Jiujun Zhang

Lithium-Ion Batteries: Fundamentals and Applications
Yuping Wu

Lead-Acid Battery Technologies: Fundamentals, Materials, and Applications
Joey Jung, Lei Zhang, and Jiujun Zhang

Solar Energy Conversion and Storage: Photochemical Modes
Suresh C. Ameta and Rakshit Ameta

Electrochemical Energy: Advanced Materials and Technologies
Pei Kang Shen, Chao-Yang Wang, San Ping Jiang, Xueliang Sun, and Jiujun Zhang

Electrolytes for Electrochemical Supercapacitors
Cheng Zhong, Yida Deng, Wenbin Hu, Daoming Sun, Xiaopeng Han, Jinli Qiao, and Jiujun Zhang

Electrochemical Reduction of Carbon Dioxide: Fundamentals and Technologies
Jinli Qiao, Yuyu Liu, and Jiujun Zhang

Metal–Air and Metal–Sulfur Batteries: Fundamentals and Applications
Vladimir Neburchilov and Jiujun Zhang

Metal–Air and Metal–Sulfur Batteries

Fundamentals and Applications

Edited by
Vladimir Neburchilov
Jiujun Zhang

CRC Press
Taylor & Francis Group
Boca Raton London New York

CRC Press is an imprint of the
Taylor & Francis Group, an **informa** business

CRC Press
Taylor & Francis Group
6000 Broken Sound Parkway NW, Suite 300
Boca Raton, FL 33487-2742

First issued in paperback 2019

ISBN-13: 978-1-4822-5853-0 (hbk)
ISBN-13: 978-0-367-87085-0 (pbk)

Library of Congress Cataloging-in-Publication Data

Names: Neburchilov, Vladimir, editor. | Zhang, Jiujun, editor.
Title: Metal-air and metal-sulfur batteries : fundamentals and applications / edited by Vladimir Neburchilov and Jiujun Zhang.
Description: Boca Raton : Taylor & Francis, CRC Press, 2016. | Series: Electrochemical energy storage and conversion | Includes bibliographical references and index.
Identifiers: LCCN 2016008934 | ISBN 9781482258530 (alk. paper)
Subjects: LCSH: Storage batteries--Materials. | Metals. | Sulfur.
Classification: LCC TK2941 .M48 2016 | DDC 621.31/2424--dc23
LC record available at http://lccn.loc.gov/2016008934

Visit the Taylor & Francis Web site at
http://www.taylorandfrancis.com

and the CRC Press Web site at
http://www.crcpress.com

Contents

Series Preface

The goal of the Electrochemical Energy Storage and Conversion book series is to provide a comprehensive coverage of the field, with topics focusing on fundamentals, technologies, applications, and latest developments, including secondary (or rechargeable) batteries, fuel cells, supercapacitors, CO_2 electroreduction to produce low-carbon fuels, electrolysis for hydrogen generation/storage, and photoelectrochemistry for water splitting to produce hydrogen, among others. Each book in this series is self-contained, written by scientists and engineers with strong academic and industrial expertise who are at the top of their fields and on the cutting edge of technology. With a broad view of various electrochemical energy conversion and storage devices, this unique book series provides essential reads for university students, scientists, and engineers and allows them to easily locate the latest information on electrochemical technology, fundamentals, and applications.

Jiujun Zhang
National Research Council of Canada
Richmond, British Columbia, Canada

Preface

Electrochemical energy storage/conversion technologies, including batteries, fuel cells, and supercapacitors, play a key role in the world movement for green and sustainable energy. These technologies can compensate for the gaps between demand and supply of electricity and be used in the transportation sector, providing reliable, low-cost, and clean power sources. With respect to this, metal–air and metal–sulfur batteries (MABs/MSBs) represent one kind of the most efficient energy storage technologies with high round trip efficiency, high energy density, fast response at peak demand/supply of electricity, and decreased weight due to the use of atmospheric oxygen as one of the main reactants. Currently, the most developed MABs/MSBs with technical potential and market perspectives are those with zinc–air, lithium–air, aluminum–air, magnesium–air, and lithium–sulfur combinations. However, both the aspects of materials and technologies still face some challenges in terms of performance, cost, and applications. With respect to this, a book covering all important areas of MAB/MSB materials and technologies as well as their applications in electrochemical energy storage and conversion should be highly useful.

This book presents an overview of the main MABs/MSBs from fundamentals to applications. The recent technological trends in their development will also be reviewed. Detailed analysis of these batteries at the material, component, and system levels allows to evaluate the different approaches of their integration. By providing a systematic overview of the components, design, and integration, this book will benefit researchers, students, industrial professionals, and manufacturers who are working in the development of MABs/MSBs.

The authors have both strong academic and technical expertise and provide fundamentals as well as up-to-date technical knowledge and information. Their contributions emphasize and systematically summarize in detail MAB/MSB science and technologies, current achievements, challenges, and future directions, making the book attractive to researchers working in the fields of clean energy. This book contains six chapters: Chapter 1 focuses on zinc–air batteries, Chapter 2 on lithium–air batteries, Chapter 3 on aluminum–air batteries, Chapter 4 on magnesium–air batteries, Chapter 5 on lithium–sulfur batteries, and Chapter 6 on vanadium–air redox flow batteries. We express our deep appreciation to all the authors who have contributed their high-quality chapters to this book. We also thank the CRC Press book editor, Allison Shatkin, for her professional assistance and strong support during this project.

Finally, we welcome any constructive comment for further improvements to this book.

Vladimir Neburchilov, PhD
Vancouver, British Columbia, Canada

Jiujun Zhang, PhD
Richmond, British Columbia, Canada

Editors

Dr. Vladimir Neburchilov earned his PhD in electrochemistry/electrocatalysis from Karpov Institute of Physical Chemistry (Moscow, Russia). After postdoctoral fellowships at the University of Alberta (Edmonton, Canada) and National Research Council of Canada Institute for Fuel Cell Innovation (NRC-IFCI) (Vancouver, Canada), he continued his scientific career as a research officer at NRC-IFCI. He has more than 27 years of experience in the development of innovative electrocatalysts for metal–air batteries (MABs), lithium–sulfur batteries, fuel cells (polymer electrolyte membrane fuel cells [PEMFCs], direct methanol and ethanol fuel cells [DMFC/DEFC], and microbial fuel cell [MFC]), and membrane chlorine electrolyzers (dimensionally stable anodes [DSAs]). Dr. Neburchilov has four diplomas, including a master of science and an MBA (SFU Beedie School of Business, Vancouver) in management of technology. He has experience in the commercialization of his own electrochemical technologies. He holds 8 international patents and has coauthored more than 32 papers, refereed journal papers, 4 book chapters, and 1 edited/coauthored book. He has prepared more than 25 industrial technical reports.

Dr. Jiujun Zhang is a principal research officer at the National Research Council of Canada (NRC), fellow of the International Society of Electrochemistry (ISE), and fellow of the Engineering Institute of Canada (EIC). The technical expertise areas of Dr. Zhang are electrochemistry, photoelectrochemistry, spectroelectrochemistry, electrocatalysis, fuel cells (PEMFC, SOFC, and DMFC), batteries, supercapacitors, and electrylysis. He earned his BS and MSc in electrochemistry from Peking University in 1982 and 1985, respectively, and his PhD in electrochemistry from Wuhan University in 1988. Starting in 1990, he carried out three terms of postdoctoral research at the California Institute of Technology, York University, and the University of British Columbia. Dr. Zhang holds more than 14 adjunct professorships, including 1 at the University of Waterloo, 1 at the University of British Columbia and 1 at Peking University. Until now, Dr. Zhang has approximately 400 publications with more than 17,000 citations, including 230 refereed journal papers with an *H-Index* of 61, 15 edited/coauthored books, 37 book chapters, 110 conference oral and keynote/invited presentations, and more than 10 US/EU/WO/JP/CA patents, and he has produced in excess of 90 industrial technical reports. Dr. Zhang serves as the editor/editorial board member for several international journals as well as editor for book series (Electrochemical Energy Storage and Conversion, CRC Press).

Contributors

Zhongwei Chen
Department of Chemical Engineering
University of Waterloo
Waterloo, Ontario, Canada

L.C. De Jonghe
PolyPlus Battery Company
Berkeley, California

Fei Ding
National Key Lab of Power Sources
Tianjin Institute of Power Sources
Tianjin, People's Republic of China

Bruce W. Downing
MagPower Systems Inc.
White Rock, British Columbia, Canada

Jing Fu
Department of Chemical Engineering
University of Waterloo
Waterloo, Ontario, Canada

Nobuyuki Imanishi
Graduate School of Engineering
Mie University
Tsu, Japan

Vladimir Neburchilov
Energy, Mining and Environment
National Research Council of Canada
Vancouver, British Colombia, Canada

E.S. Nimon
PolyPlus Battery Company
Berkeley, California

P.N. Ross
PolyPlus Battery Company
Berkeley, California

S.J. Visco
PolyPlus Battery Company
Berkeley, California

Sihui Wang
National Key Lab of Power Sources
Tianjin Institute of Power Sources
Tianjin, People's Republic of China

Osamu Yamamoto
Graduate School of Engineering
Mie University
Tsu, Japan

Jiujun Zhang
Energy, Mining and Environment
National Research Council of Canada
Vancouver, British Columbia, Canada

Qingqing Zhang
National Key Lab of Power Sources
Tianjin Institute of Power Sources
Tianjin, People's Republic of China

Qing Zhao
National Key Lab of Power Sources
Tianjin Institute of Power Sources
Tianjin, People's Republic of China

Hai Zhong
National Key Lab of Power Sources
Tianjin Institute of Power Sources
Tianjin, People's Republic of China

Jun Zong
National Key Lab of Power Sources
Tianjin Institute of Power Sources
Tianjin, People's Republic of China

1 Zinc–Air Batteries
Fundamentals and Applications

Jing Fu and Zhongwei Chen

CONTENTS

1.1 INTRODUCTION

Electrical energy plays an important role in our everyday life as it can be stored and applied for general energy demands. Various types of electrochemical power sources are available for employing chemical compounds for conversion media and making use of redox reactions to yield or to store electrical energy. Generally, these power sources are divided into two types: energy storage (e.g., batteries and supercapacitors) and conversion (e.g., solar cell) systems. Considerable effort has been directed toward advanced technologies for more efficient energy storage systems in a wide range of applications, from microelectrochemical devices to portable electronics and even to transportation and grid storage. Among viable alternatives, batteries are preferable to fuel cells for specific applications that require moderate power density, and to supercapacitors applications that require higher energy density.

Currently, the most widely available batteries are lead–acid batteries, nickel–cadmium, batteries, and lithium-ion batteries for applications such as portable devices and electric vehicles (EVs). There has been some exciting progress, especially in the field of lithium-ion batteries; however, their practical energy densities

TABLE 1.1

Comparison of Different Types of Metal–Air Batteries

Battery Systems	Fe–Air	Zn–Air	Al–Air	Mg–Air	Na–Air	K–Air	Li–Air
Year invented	1968	1878	1962	1966	2012	2013	1996
Cost of metals ($/kg)	0.40	1.85	1.75	2.75	1.7	~20	68
Theoretical voltage (V)	1.28	1.65	2.71	3.09	2.27	2.48	2.96
Theoretical energy density (Wh/kg)	763	1086	2796	2840	1106	935	3458
Electrolyte for practical batteries	Alkaline	Alkaline	Alkaline or saline	Saline	Aprotic	Aprotic	Aprotic
Practical voltage (V)	~1.0	1.0–1.2	1.1–1.4	1.2–1.4	~2.2	~2.4	~2.6
Practical energy density (Wh/kg)	60–80	350–500	300–500	400–700	Unclear	Unclear	Unclear
Primary (P) or electrically rechargeable (R)	R	R	P	P	R	R	R

Source: Li, Y. et al., *Chem. Soc. Rev.*, 43(15), 5257, 2014.

are still less than sufficient for the extended range of EV propulsion; as well, they are expensive and potentially not safe. On the other hand, for energy storage, metal–air battery systems as candidates enhance energy density, efficiency, cost, and safety compared to lithium-ion batteries. The primary/secondary metal–air batteries such as zinc (Zn)–, aluminum (Al)–, lithium (Li)–, iron (Fe)–, and magnesium (Mg)–air batteries have been paid a great deal of attention. The comparison of different types of metal–air batteries is summarized in Table 1.1.[1] Al–air and Mg–air batteries have high theoretical energy densities and working voltages; however, their practically attainable values are much lower due to the parasitic corrosion reaction with hydrogen evolution. Moreover, they are not electrically rechargeable since the electrodeposition of Al and Mg is not thermodynamically feasible in aqueous electrolytes. Fe–air batteries can be electrically recharged and are capable of a long cycle life (>1000 cycles); however, their energy density is the lowest compared with other metal–air batteries. Nonaqueous Li–air batteries are appealing due to their highest theoretical energy density. However, they are plagued by intrinsic performance limitations (low-power capability and poor cyclability). Compared with their aqueous counterparts (Al–, Mg–, Fe–, and Zn–air), Li–air batteries have a starkly different battery electrochemistry. Oxygen reduction reaction (ORR) in nonaqueous solvents proceeds at a rate orders of magnitude slower than in aqueous electrolytes. This leads to the formation of insoluble metal peroxide or superoxide particles, the accumulation of which at the air electrode blocks oxygen diffusion, and gradually shuts off battery reactions. This characteristic essentially eliminates the feasibility of mechanical recharging in nonaqueous metal–air batteries unless a method to dissolve the discharge product can be identified.

Zinc–air batteries (ZABs) have technological advantages over other metal–air battery systems due to the relatively high stability and reversibility of zinc material, especially in alkaline electrolytes, low cost, and safety. Thus, they are more amenable to be engineered into practical small-scale (portable and wearable devices) and large-scale (VE and stationary grid storage) applications. Commercialization of the low-rate primary ZAB started in 1932, and they are used only in hearing aids, watches, navigation lights, and railway signals. However, problems with capacity retention and cycle life, which are related mainly to zinc dendrite formation, shape change during zinc regeneration, and unsatisfactory air electrode performance, have slowed down the commercialization of the secondary (rechargeable) ZAB.[2] Recently, several long-life rechargeable zinc–air flow batteries developed at Eos Energy Storage and ZincNyx Energy Solution, Inc. are anticipated for a wide range of applications, such as VEs and grids.

This chapter provides a comprehensive understanding of the fundamental principles, technologies, and applications of ZABs.

1.2 GENERAL CHARACTERISTIC

ZABs have a theoretical cell voltage of 1.65 V with a high gravimetric energy density of 1350 Wh/kg based on zinc active material.[3] However, the polarization effects resulting mainly from the activation, the ohmic, and the concentration loss of an air electrode degrade the battery voltage and performance under practical operating conditions. For example, the discharging voltage of ZAB generally ranges from 1.0 to 1.3 V, whereas the charging voltage is approximately 1.8 V or higher. These high overpotentials during the discharge–charge process result in a loss of overall energy efficiency and performance even before any other factors are considered. During discharge, ZAB function as power generation through the electrochemical coupling of a zinc anode to the air cathode in the presence of an electrolyte with an inexhaustible cathode reactant (oxygen) from the atmosphere. The electrons liberated at the zinc anode travel through an external load to the air cathode, producing zinc ions. At the same time, atmospheric oxygen diffuses into the porous air electrode and is ready to be reduced to hydroxide ions via ORR (forward reaction 1.1) at a three-phase reaction site, which is the interface of oxygen (gas), electrolytes (liquid), and electrocatalysts (solid).[4] The generated hydroxide ions then migrate from the reaction site to the zinc electrode forming zincate ions (forward reaction 1.2), which then further decompose to insoluble zinc oxide (ZnO) at supersaturated zincate concentrations (forward reaction 1.3). Equation 1.4 shows the overall anode reaction. During charge, ZABs are capable of storing electric energy through a reversed redox reaction occurring in the oxygen evolution reaction (OER) (backward reaction 1.1) at the electrode–electrolyte interface, whereas zinc is deposited at the cathode surface (backward reaction 1.4). The overall discharge and charge reactions in an alkaline electrolyte are the following.

The air electrode reaction:

$$O_2 + 2H_2O + 4e^- \Leftrightarrow 4OH^-, \quad E = -0.40\,V\,vs.\ SHE \qquad (1.1)$$

The zinc electrode reactions:

$$Zn + 4OH^- \Leftrightarrow Zn(OH)_4^{2-} + 2e^-, \quad E = +1.25 \text{ V vs. SHE} \tag{1.2}$$

$$Zn(OH)_4^{2-} \Leftrightarrow ZnO + 2OH^- + H_2O \tag{1.3}$$

$$Zn + 2OH^- \Leftrightarrow ZnO + H_2O + 2e^-, \quad E = +1.25 \text{ V vs. SHE} \tag{1.4}$$

The overall reaction:

$$2Zn + O_2 \Leftrightarrow 2ZnO, \quad E = 1.65 \text{ V vs. SHE} \tag{1.5}$$

A schematic of a rechargeable ZAB and a picture of typical ZAB components are shown in Figures 1.1 and 1.2, respectively.[4]

Note that zinc is thermodynamically unstable in an aqueous electrolyte because its reduction potential lies well above that of hydrogen electrodes. Thus, there is

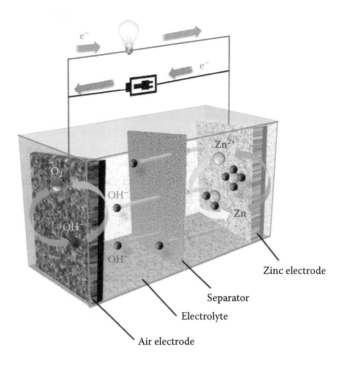

FIGURE 1.1 A schematic of rechargeable zinc–air battery during the discharge and charge process.

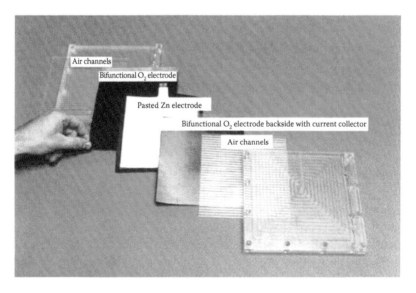

FIGURE 1.2 Components of a typical zinc–air battery. Pasted zinc electrode (white) is located between two binational oxygen electrodes and two corrugated Plexiglas plates for air supply. (From Hass, O. and Wesemael, J.V., *Encyclopedia of Electrochemical Power Sources*, 2009, p. 384.)

always a driving force that favors the corrosion or oxidation of zinc along with the evolution of hydrogen, as expressed by the following reaction 1.6:

$$Zn + 2H_2O \Leftrightarrow ZnO + H_2 \text{ (self-discharge)} \qquad (1.6)$$

Hydrogen evolution reaction (HER) is detrimental to the safety of the ZAB system due to the buildup of gas pressure inside the battery. Moreover, this parasitic corrosion reaction, or so-called self-discharge, degrades the columbic efficiency of the zinc electrode and must be inhibited to minimize the loss of capacity.

Zinc passivation by forming ZnO film on the electrode surface during discharge is another problem that prevents the maximum utilization of zinc active materials, resulting in the loss of specific energy. The detailed mechanisms of zinc passivation in the alkaline solutions are rather complex.[5] The zinc is initially oxidized to form zinc hydroxide ($Zn(OH)_2$) when treated anodically in the alkaline solution. The $Zn(OH)_2$ then dissolves in the electrolyte to produce $Zn(OH)_4^{-2}$ or $Zn(OH)_3^{-}$ compounds, and if the anodic process is continued long enough, a solid ZnO film begins to precipitate onto the electrode surface.[5,6]

ZAB can be categorized into various types depending on configuration, electrolytes, and electrodes (Table 1.2). The mechanically rechargeable designs compromise the exchange of a "fresh" zinc electrode and the refueling of a "fresh" zinc slurry/electrolyte during a charge mode, whereas the electrical recharging of ZAB requires that either a third separate electrode (to sustain OER for charge) or a bifunctional electrode (a single electrode capable of both ORR and OER) be recharged inside the batteries.

TABLE 1.2

Zinc–Air Battery Category

Category	Battery Type
Configurations	(i) The primary type
	(ii) The electrically rechargeable type (bifunctional air electrode)
	(iii) The hydraulically rechargeable type (flow system)
Electrolytes	(i) Basic/acidic/neutral-aqueous electrolytic type
	(ii) Ionic liquid-nonaqueous electrolytic type
	(iii) Solid-state electrolytic type

The measured terminal battery voltage appearing at any particular discharge state depends on the load current rate, the crystal structure of the electrode components, the internal impedance inside the battery, and the state of charge. Note that each battery system has its own characteristic actual voltage and discharge curve. For example, certain batteries such as lead–acid batteries have a pronounced discharge slope, whereas others such as lithium-ion batteries and nickel–cadmium batteries have fairly constant discharge profiles, with abrupt falloffs in voltage at the end of discharge. A sloping discharge characteristic causes the power and energy output to decline progressively over the discharge cycle and thus gives rise to problems for some applications that need a stable output supply. However, flat discharge curves are generally preferred because of the maximized energy output. The flat voltage characteristic could be the single-phase region of a reactant with a wide range of stoichiometry or the equilibrium between two or more phases.[7] Typically, ZABs maintain a relatively flat output voltage throughout the discharge cycle at constant current drains, as shown in Figure 1.3. Note that the "knee" (the sharp voltage inflection) of

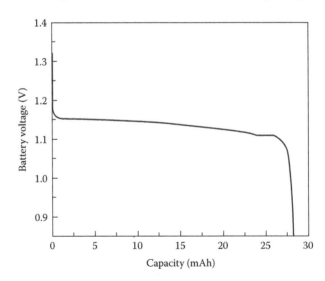

FIGURE 1.3   Typical discharge profile for zinc–air batteries.

the voltage at the end of discharge indicates a large internal resistance of ZnO, resulting in large polarization effect.

1.3 ELECTRODE

1.3.1 ZINC ELECTRODE

Zinc electrodes are utilized in a vast majority of the primary ZAB, but have found rather limited use in secondary (rechargeable) batteries. The redistribution of zinc active material (namely shape change) and formation of zinc dendrites are the main issues that limit the performance and lifetime of ZAB when subjected to a number of discharge–charge cycles. These morphology problems are related to two important characteristics of zinc: (1) its fast electrochemical kinetics in alkaline electrolytes and (2) its high solubility as in the form of zincate ions, which can freely migrate into the electrolytes and can redeposit zinc away from where it was consumed. As a result, the zinc deposits become unevenly distributed over the electrode surface or trigger the growth of dendrites, which gradually decrease battery performance, or even more seriously, cause a short circuit inside batteries.

1.3.1.1 Zinc Dendrites and Shape Change

Generally, the morphology of zinc deposits during charge can be mossy or spongy (at low current densities), layer-like (at middle current densities), granular, dendritic, or clustered (at high current densities) in form, as shown in Figure 1.4.[8] The regenerated

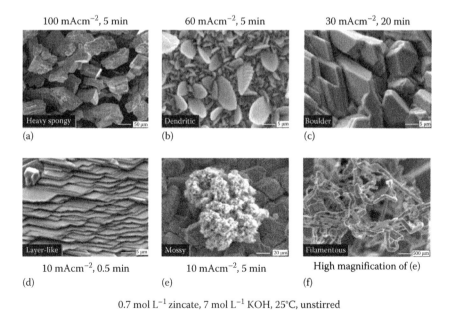

$100\ mAcm^{-2}$, 5 min $60\ mAcm^{-2}$, 5 min $30\ mAcm^{-2}$, 20 min

(a)   (b)   (c)

$10\ mAcm^{-2}$, 0.5 min $10\ mAcm^{-2}$, 5 min   High magnification of (e)

(d)   (e)   (f)

$0.7\ mol\ L^{-1}$ zincate, $7\ mol\ L^{-1}$ KOH, 25°C, unstirred

FIGURE 1.4 Different types of deposit morphology obtained from KOH electrolyte: (a) clustered, (b) dendrite, (c) boulder, (d) layer types, (e) mossy, and (f) filament. (From Zhang, X., *Encyclopedia of Electrochemical Power Sources*, 2009, p. 454.)

zinc tends to occur especially at corners, edges, or protrusions where the current density is locally increased.[9]

There are two models: activation- and diffusion-controlled dendrite tip propagation for zinc electrodeposition in alkaline electrolytes. Activation-controlled deposition plays a leading role at low current densities due to ions translating into atoms, overcoming the activation energy barrier, whereas diffusion-controlled deposition dominates at high current densities.[10] In practical applications, charging batteries at a high current density (fast charge) is always desired due to the considerations of time, cost, and performance. However, high overpotentials prevalent at high charging rates favor high nucleation rates of zinc deposition, thereby approaching the mass transport limited rate (e.g., the limiting current density) and producing needlelike dendrites.[11]

Shape change refers to the morphology change involving with alteration of the zinc electrode geometric area, where zinc active material leaves one location and agglomerates in others. This phenomenon is mainly caused by differences in the current density distribution over different zones of the zinc electrode during discharge–charge cycling. Two established models describe the zinc electrode shape change mechanisms: the concentration cell model[12] and the electrolyte flow model.[13]

1.3.1.2   Technologies

Various physical techniques that optimize battery systems and operating conditions have attempted to eliminate the problem of zinc electrode recharging. One method is to mechanically replace the electrode with fresh zinc after each full discharge. This mechanically rechargeable system has been developed for large power source applications. Hydrodynamic refueling is another effective way to overcome the issue of dendritic growth. In this system (Figure 1.5), the battery is operated as a zinc–air fuel cell, where the zinc slurry in the form of particles or dendrites is continually supplied through a flowing electrolyte in the cell.[8] As in quiescent electrolyte system, the gradient of zincate ion concentration between the bulk electrolyte and the interfacial layer near the electrode increases as a result of electrochemical reaction. This increase affects the zinc deposition morphology due to the changes of zincate

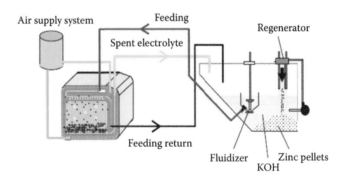

FIGURE 1.5   Schematic illustration of the flow of material in a zinc–air fuel cell system. (From Zhang, X., *Encyclopedia of Electrochemical Power Sources*, 2009, p. 454.)

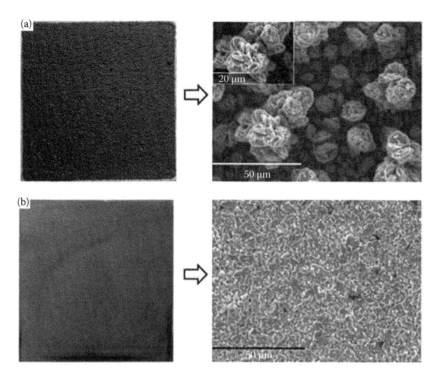

FIGURE 1.6 Deposited zinc morphologies in weight 40% KOH with 1 M zinc oxide with electric quantity for 10 min at the current density of 100 mA/cm². (a) Quiescent electrolyte; (b) flowing electrolyte. (From Wang, K. et al., *J. Power Sources*, 271, 65, 2014.)

concentration at different reaction zones. Thus, to uniformly concentrate reactive species of the solutions, it is necessary to make the electrolyte solution flow. Figure 1.6 shows that two different zinc depositing morphologies are obtained under the conditions of the quiescent electrolyte and the flowing electrolytes.[14] The electrolyte flow not only reduces concentration gradients between the bulk of the electrolyte solution and surface electrolyte near the zinc electrode, but helps to improve zincate ion migration and suppress zinc dendritic growth. It also takes away gas bubbles in the ZAB with alkaline electrolytes.

In addition, appropriate charging mode make suppressing dendritic growth possible. Pulsating current or periodic reverse current deposition is conducive to improve depositing morphology due to zincate ions having enough diffusion time. Pulsating currents, namely pulse current with different duty cycles, can improve the morphology of deposited zinc. Arouete et al.[15] experimented with pulsed-current electrodeposition for surface smoothing. Despic[16] showed zinc dendrite suppression using high-frequency square-wave potential pulses.

Some researchers have investigated how additives in the electrolytes affect the morphology of regenerated zinc. The use of ppm levels of additives,[17] such as polyethylenimine (PEI) and polyethylene glycol (PEG), in the electrolytes has been approved to alleviate the shape change and suppress the formation of dendrites to

1 mm

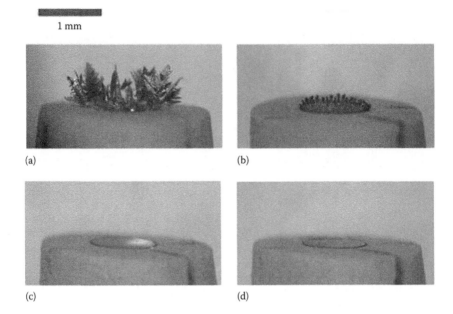

(a)   (b)

(c)   (d)

FIGURE 1.7   In situ optical microscopy of a Zn wire electrode after potentiostatic Zn electrodeposition at −1.57 V (vs. Hg/HgO) for 30 min. Electrolyte contains 0.1 M ZnO + 4.0 M KOH with various concentrations of PEI: 0 ppm (a), 10 ppm (b), 50 ppm (c), and 100 ppm (d). At a PEI concentration of 50 ppm or higher, dendritic morphology during Zn plating is eliminated. (From Banik, S. et al., *Electrochim. Acta*, 179, 475, 2015.)

some extent. Stephen et al.[18] investigated the effect of PEI concentration on zinc dendrite growth suppression and found that increasing the content of PEI from 10 to 100 ppm caused significant suppression, as shown in Figure 1.7. The suppression efficacy of additives may occur because the adsorption of the additives on the zinc surface lowers the kinetic parameters, for example, the exchange current density (i_0), during the process of Zn regeneration. This adsorption retards deposition at an activation-controlled dendrite tip, thus producing dendrite suppression.

Calcium hydroxide is also commonly used as an additive to the electrolyte solution to reduce the solubility of the zincate ions. The presence of calcium hydroxide leads to a relatively insoluble compound of $Ca(OH)_2 \cdot Zn(OH)_4 \cdot 2H_2O$, which can insignificantly improve the shape change of the zinc electrode. Moreover, the zinc can easily be redeposited onto the electrode surface during recharge since the reaction is reversible.

$$2Zn(OH)_4^{2-} + Ca(OH)_2 + 2H_2O \Leftrightarrow Ca(OH)_2 \cdot 2Zn(OH)_2 \cdot 2H_2O + 4OH^-$$

The shape change and dendrite formation can also be effectively curtailed if the free electrolyte volume between and around the electrode is minimized and the separator is firmly pressed on the electrodes. Kohei[19] demonstrated that a high efficient

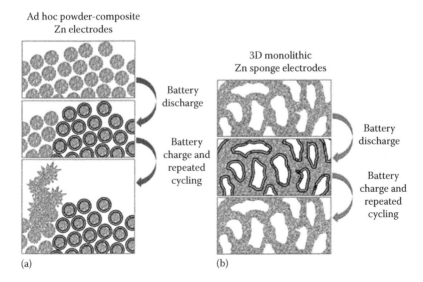

Ad hoc powder-composite
Zn electrodes

3D monolithic
Zn sponge electrodes

Battery
discharge

Battery
charge and
repeated
cycling

Battery
discharge

Battery
charge and
repeated
cycling

(a)   (b)

FIGURE 1.8 Cross-sectional schematic of the dissolution–precipitation of ZnO in (a) conventional ad hoc powder-bed electrodes and (b) the 3D-wired Zn sponge electrode. Interparticle connectivity is lost in powder-composite electrodes leading to regions of high local current density and dendrite formation. The 3D Zn@ZnO core–shell architecture is maintained throughout charge–discharge, leading to high Zn utilization, controlled ZnO deposition within the void space, and diminished shape change upon cycling. (From Parker, J. et al., *Energy Environ. Sci.*, 7(3), 1117, 2014.)

charge (less dendrite formation) can be achieved by employing an anion-exchange ionomer (AEI)-modified ZnO electrode. During the discharging processes, the AEI film on top of the electrode surface hindered the transport of zincate anions from the vicinity of zinc electrodes and formed a region of condensed zincate anions. During charge, zinc was reductively formed from the condensed zincate anions, and the consumed zincate anions were reproduced from ZnO, which mitigated the zinc growth of dendrites.

Additionally, the porous structure of the zinc electrode itself exhibits an enhanced capability of preventing dendrite growth. Joseph proposed a 3D-wired Zn architecture for primary ZAB that innately suppress dendrite formation, as shown in Figure 1.8.[20]

1.3.2 AIR ELECTRODE

The development of a reliable bifunctional air electrode, where consumption and evolution of oxygen (so-called ORR and OER) take place, is quite a challenge for rechargeable ZAB. Both reactions show considerable high overpotentials. The operating range of a bifunctional air electrode spans from 0.6 to 0.7 V (RHE) during the ORR (discharge mode) to more than 1.8 V (RHE) during the OER (charge mode).[21] The utilization of two separate electrodes for ORR and OER has been proposed;

however, it adds to the weight and complexity of the cell system. It is therefore advantageous to use the bifunctional air electrode that allows ORR and OER occurring at the same electrode–electrolyte interface.

1.3.2.1   Oxygen Reduction Reaction

The ORR occurring at the ZAB during discharge is similar to the ORR in alkaline hydrogen fuel cell systems with the hydroxide ions being the main product. Because of this similarity, some ORR catalysts for alkaline fuel cells are also promising candidates for the ZAB.[1] N. A. Anastasijevic comprehensively investigated the ORR pathways in alkaline electrolytes, as shown in Figure 1.9.[22] The ORR in aqueous alkaline solutions can proceed through parallel pathways, one leading to OH− through a direct four-electron pathway and the other one leading to peroxide through a two-electron pathway. The direct four-electron reduction pathway generally is predominant on noble metal catalysts such as platinum and silver, and metal nitrides, and some transition metal oxides having the perovskite or pyrochlore structure. The peroxide path of ORR is more common and is favored on graphite and most other carbons, some transition metal oxides, and mixed oxides of the transition metals.[23] For example, carbon can be used as a catalyst for ORR in alkaline electrolytes, although the electrode polarization is high due to the formation of peroxide.[24] The perovskite catalysts such as $La_{0.6}Ca_{0.4}CoO_3$ have been widely investigated for ZAB applications due to their high catalytic activity for oxygen reduction.[25]

1.3.2.2   Oxygen Evolution Reaction

A new look should be taken at the catalysts for OER. Besides being catalytically active, the catalysts for OER should be endurable and conductive in the presence of a highly reactive form of oxygen. Platinum is a rather poor electrocatalyst for OER due to the formation of a stable and low electrical-conductive oxide layer. In fact, platinum has been observed to dissolve during charge and form a finely divided platinum black deposit on the zinc electrode, resulting in low charging efficiency of the ZAB.[26] Ruthenium, although being a good electrocatalyst for OER, is strongly corroded at the potential of oxygen evolution. On the contrary, electrical-conductive noble metal oxides such as ruthenium oxide (RuO_2) or iridium dioxide (IrO_2) show rather good OER performance. Some nonprecious metals and transition metal oxides also show good performance as the effective electrocatalysts for OER in alkaline electrolytes. For instance, nickel as a low-cost catalyst exhibits good OER activity and stability in alkaline electrolytes. The spinel-type oxides such as nickel cobalt oxide ($NiCo_2O_4$) or copper cobalt oxide ($CuCo_2O_4$) also show promising OER performance. Obviously, the electronic properties determined by the d-electrons, the composition, and the stoichiometry of the metal oxide, as well as the vacancies present in the metal oxides, are important factors for the OER. Also, geometric factors in the catalyst lattice are of importance.[22]

During charge process, the generated oxygen is highly reactive, which causes a severely corrosive condition to the catalyst support materials, particularly the high-surface-area carbon materials. Thus, it is of utmost importance to prevent the corrosion of the catalyst support either by the use of highly corrosion-resistant carbon

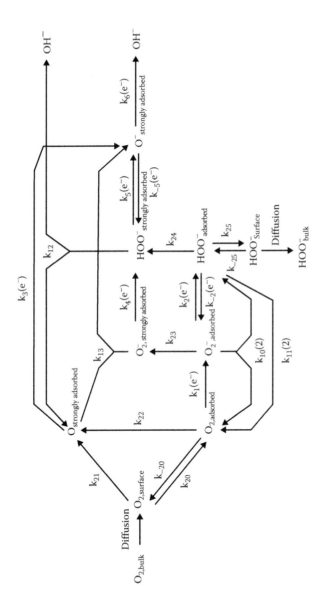

FIGURE 1.9 Reaction mechanisms of the oxygen reduction in alkaline electrolytes. (From Jöerissen, L., *Encyclopedia of Electrochemical Power Sources*, 2009, p. 356.)

materials, such as graphitized carbons, or by the use of stable metallic substrates, such as niobium-doped titanium dioxide.

1.3.2.3 Design of Bifunctional Air Electrode

The idea of using porous gas diffusion layer (GDL) in alkaline fuel cells can be applied to the air electrode design for ZAB. Often, the GDL is a double-layered structure—a hydrophilic catalytic layer near the electrolyte and a hydrophobic layer having fine porosity for gas transport. The porous structure of the GDL involving high-surface area catalysts must be optimized to increase the reaction sites at the three-phase boundary between gas, catalyst, and electrolyte, achieving significant current densities.

Since the ORR and OER can be carried out at nonprecious metal catalysts by using the alkaline electrolytes, a wide variety of transition metal oxides and mixed transition metal oxides have been successfully used as bifunctional catalysts for use in rechargeable ZAB. Three types with the spinel, the perovskite, and the pyrochlore structure are predominant. Among them, the spinel-type catalysts are particularly interesting for efficient bifunctional catalysis due to their relatively high bifunctionality, excellent electrochemical durability, and low electrical resistance. Spinels are a group of oxides with the formula AB_2O_4, where A is a divalent metal ion (such as Mg, Fe, Ni, Mn, or Zn) and B is a trivalent metal ion (such as Al, Fe, Cr, or Mn).[23] For instance, spinel-type Co_3O_4 has been extensively investigated as an active and durable transition metal oxide catalyst to minimize the cost and weight of electrically rechargeable ZABs.[27] However, the performance of transition metal oxides and mixed transition metal oxides is often limited by their low bulk conductivity and limitations in the active surface area. Those limitations can be extended by chemical synthesis methods, such as sol-gel and coprecipitation methods, or by tailoring the catalyst composition. Generally, the selection of an appropriate electrocatalyst involves a compromise between electrocatalytic activity, thermodynamic stability, corrosion resistance, fabrication and materials cost, and long-term stability. These properties can be tuned by changing the composition and geometry of the catalyst. The composition is directly related to electronic properties of the catalyst, which determines the strength of surface-intermediate bonds, hence the electrocatalytic activity, while the geometry is related to the actual surface areas, and active site density, which can be controlled by the catalyst morphology.

1.4 ELECTROLYTE

The electrolytes, which provide the bridge for ions migration between electrodes, should (1) be chemically inert toward the electrode materials under open circuit condition, (2) be electrochemically stable during the operation over a wide range of power demands and temperature, and (3) have a high specific ionic conductivity and low viscosity to ensure appropriate ion transport, as well as to minimize the ion concentration gradient.

Typically, aqueous alkaline solutions are utilized as electrolytes in ZAB due to the enhanced performance of the ORR/OER in the alkaline environment. Potassium hydroxide (KOH) solutions (15–33 wt%) are most widely used in ZAB as they exhibit superior ionic conductivity and kinetic behavior compared to sodium hydroxide (NaOH). However, it is well known that the drawback of using alkaline electrolytes is the formation of carbonate species due to their high solubility for carbon dioxide (CO_2). The CO_2 from the atmosphere diffuses into the cell from air cathode and consumes the electrolytes by forming the carbonate crystals, which decrease the molarity of hydroxide ions, leading to the passivation of zinc anode, and irreversibly impede air access by precipitating carbonate in the pores of air cathode. Thus, to prevent this, some researches look into neutral electrolytes as more robust alternatives with respect to traditional alkaline electrolytes for the ZAB. A type of near-neutral electrolytes based on zinc chloride and ammonium chloride was reported for rechargeable ZAB applications.[28] The prototyped cell tests proved that this type of chloride electrolyte system can sustain more than 1000 h and hundreds of discharge–charge cycles without dendrite formation under discharge–charge capacity ranging from 20 to 120 mAh.

Another issue of using aqueous electrolytes in ZAB is water evaporation due to the open system of the air electrode. On the one hand, the water loss causes the electrolytes to concentrate, and thus increase the transfer resistance of hydroxyl ions. On the other hand, insufficient water is not facilitating to support the destruction of peroxide and subsequent hydroxyl formation during ORR. Therefore, recent studies have been investigating the possibility of replacing the current caustic aqueous electrolytes (KOH), with the alternatives of nonaqueous electrolytes, such as solid-state polymer electrolytes, ionic liquids, and ceramic/glassy electrolytes.

Among those candidates, the solid-state polymer electrolytes (membranes) have drawn great attention due to their functional combination of the electrolytes and separators, resulting in more compact structure compared to the conventional aqueous ZAB. In the solid-state ZAB system, the polymer electrolysis joined directly and positioned between the zinc and air electrode. A simple approach to solid-state polymer electrolyte consists in "gelling" an alkaline electrolyte (typical KOH) with a suitable polymer matrix, such as sodium carboxymethylcellulose (CMC), poly(vinyl alcohol) (PVA), hydroponics gel and agar, and poly(acrylic acid). The ionic conductivity of some gelled electrolytes is appreciable at room temperature.[29]

Another strategy is to employ alkaline anion-exchange membranes (AEMs), a type of homogeneous membranes consisting exclusively of the cationic (i.e., ammonium) groups and the mobile counter ions (hydroxyl groups), as nonaqueous electrolyte candidates for the ZAB. Those hydroxyl ion-conducting polymers have been developed widely in alkaline fuel cells.

Room temperature ionic liquids (RTILs), which are molten salts generally composed of large organic cations and organic/inorganic anions, possess many advantages such as low volatility, nonflammability, high thermal stability, wide electrochemical window, and the capability of modifying metal electrodepositions. These important characteristics make RTILs attractive as potential electrolyte alternatives to avoid some issues noted for rechargeable ZAB applications. For example, RTILs can prevent zinc dendrite formation at the electrode surface, achieve a longer cycle life as it slows down the drying out of the electrolyte from water evaporation, suppress the self-discharge of zinc, and eliminate carbonation. Some RTILs with imidazolium and pyrrolidinium cations, together with trifluoromethanesulfonylimide (TFSI-) and dicyanamide (DCA) anions, are among the most promising candidates for use in batteries.

1.5   ELECTRIC VEHICLE APPLICATIONS

ZABs have been developed for uses in hearing aids, military field electronics, cellular phones, and laptop computers. Aside from consumer electronics, their applications for various forms of transportations and stationary power (smart grid) storage are currently under development as well. Among those applications, ZAB technologies have attracted significant interest for the EV revolution. The overriding reason for this is that ZABs provide the highest practical energy density at the lowest cost of any known battery technologies. It is well known that the main restriction for the wide popularization of EVs is their limited range. The range is governed by how much electrical energy the battery can store and the weight and the volume of the battery. Therefore, it is of utmost important for EV batteries to reach a high energy density (Wh/kg or Wh/L). Among the existing battery technologies, such as lithium-ion batteries, lead–acid batteries, nickel–zinc batteries, and nickel–metal hybrid batteries, the ZAB can provide the highest energy density. For instance, the battery that is recharged by mechanically replacing the zinc electrode can deliver a specific energy of 220 Wh/kg. In that regard, ZABs owe the superior energy density mainly to their unique "half-open" system in which the zinc reacts with weightless reactant of oxygen drawn from environment. With this advantage, the vehicles do not have to carry more "fuel" on board in the battery, greatly reducing its weight.

Another biggest advantage of using ZAB technologies in the application of EVs is their reduced safety issues. Note that there are still significant safety concerns with lithium-ion batteries, resulting in carrying much more extensive electronic monitoring and control to keep the batteries within safe operating limits. Moreover, the incidence of short circuit in the batteries, for example, during the event of a car crash, makes them discharge very quickly and become extremely hot. This can set the vehicles on fire and even explode fiercely. However, the ZAB system has its inherently short-circuit proof, shutting down the battery since the chemical reaction for discharge cannot take place. In addition, physically shutting off the air supply through a controlling system in the vehicles can inhibit the reaction as well. Therefore, ZAB is by far the safest among all the battery technologies.

ZABs have the highest specific energy and probably will be the least expensive of any EV rechargeable battery technology on the market and therefore are well suited for not only passenger cars but also commercial vehicles, buses, and industrial vehicles that have a unique combination of high daily usage, low power requirement, and in-place service infrastructure.

There are three main types of ZAB system for EV: (1) the onboard discharge-only battery pack, (2) rechargeable battery pack by mechanical exchanging of zinc electrode, and (3) rechargeable battery pack by hydraulic refueling of zinc slurry/electrolyte.

Commercial interest in ZAB for powering EVs has grown greatly since the 1990s. An Israeli company in 1994 demonstrated a ZAB-powered van that had a range of more than 420 km with highway acceleration and sustained speeds. However, to make the battery refuelable, the spent zinc electrodes have to be reconstructed. In April 1997, a Mercedes-Benz 410 van powered by a 150 kWh ZAB system drove over 424 km from Bremen to Bonn, Germany, on a single charge. Later in November 1997, the same vehicle drove 439 km from central London to central Paris on a single charge with temperatures as low as 1°C.[30]

Dreisbach Electromotive, Inc. developed a second-generation ZAB system to power Chrysler electric minivans for Southern California Edison. The vehicles are equipped with a 60 kWh ZAB in conjunction with a 10 kWh NiCd load-leveling battery providing a peak power of 104.4 kW (140 hp, 120 kW electrical peak) and over a 322 km (200 mile) range at 64 km/h (40 mph) while maintaining a 454 kg (100 lb) payload capability.[31]

Zinc–air fuel cell system from Electric Fuel Ltd. makes it possible to use clean electricity to power EVs on the road. The onboard zinc–air fuel cell modules yield a practical specific energy of around 200 Wh/kg and specific peak power of 90 W/kg at 80% depth of discharge. When the zinc–air fuel cell is in operation, oxygen is extracted from the air by electrochemically reducing it at the air cathode to hydroxide ions. These hydroxide ions then react with the zinc fuel inside the cell, forming the zinc oxide.

A similar rechargeable zinc–air cell, operating at room temperature, was being developed for use in EVs. The cell uses a planar bipolar configuration. The negative electrode consists of zinc particles in a paste form, similar to the electrode used in alkaline–manganese dioxide primary cells. The bifunctional air electrode consists of a membrane of carbon and plastic with appropriate catalysts. The electrolyte is potassium hydroxide with gelling agents and fibrous absorbing materials. A typical cell is rated at 100 Ah with an average operating voltage of 1.2 V. Specific energies up to 180 Wh/kg at the 5–10 h discharge rates and a battery life of about 1500 h have been achieved. Technical limitations are limited power density and a relatively short separator life. The air must be managed to remove carbon dioxide and to provide humidity and thermal management.

A refueling ZAB system developed at Lawrence Livermore National Laboratory combines atmospheric oxygen and zinc pellets in alkaline electrolytes to generate power with by-products of zinc oxide. The battery weighs only one-sixth as much as standard lead–acid batteries and occupies one-third the space to operate. In addition, because the battery is easily refuelable by replacing spent electrolyte with a fresh one, it promises trouble-free, nearly 24-h-a-day operation for numerous kinds of EVs, from forklifts to delivery vans and passenger cars.[32]

More recently, a long-life zinc–air flow battery developed by Eos Energy Storage is anticipated for EVs and grid applications. The energy density of this battery is expected to be around 100 Wh/kg, and 400 Wh/L electrolyte can be removed and saved for reconditioning. A Nissan Leaf integrated with a 100 kWh zinc–air flow battery has a range up to 340 miles.

REFERENCES

1. Y. Li, H. Dai, Recent advances in zinc-air batteries, *Chem. Soc. Rev.*, 43 (2014), 5257–5275.
2. T.B. Reddy, *Linden's Handbook of Batteries*, 4th edn., McGraw-Hill Education, New York (2011).
3. J. Goldstein, I. Brown, B. Koret, New developments in the electric fuel ltd. zinc/air system, *J. Power Sources*, 80 (1999), 171–179.
4. O. Hass, J. V. Wesemael, Secondary batteries – metal air systems: Zinc-air: Electrical recharge, *Encyclopedia of Electrochemical Power Sources*, (2009), 384–392. doi:10.1016/B978-044452745-5.00169-6.
5. T.P. Dirkse, The behavior of the zinc electrode in alkaline solutions: V. Supersaturated zincate solutions, *J. Electrochem. Soc.*, 128 (1981), 1412–1415.
6. T.P. Dirkse, The behavior of the zinc electrode in alkaline solutions: II. Reaction orders at the equilibrium potential, *J. Electrochem. Soc.*, 126 (1979), 541–543; T.P. Dirkse, The behavior of the zinc electrode in alkaline solutions: IV. The effect of ionic strength in the Tafel region, *J. Electrochem. Soc.*, 127 (1980), 1452–1456.
7. C.S.C. Sequeira, A. Hooper, *Solid State Batteries*, Springer, Dordrecht, Netherlands (1985).
8. X.G. Zhang, Secondary batteries – zinc systems: Zinc electrodes: Overview, *Encyclopedia of Electrochemical Power Sources*, (2009), 454–468. doi:10.1016/B978-044452745-5.00166-0.
9. I. Kim, R. Weil, An electron miscroscopy study of the initial stages of dendrite formation in electrodeposited zinc, *Surf. Technol.*, 25 (1985), 1–6.
10. J.W. Diggle, A.R. Despic, J.O.M. Bockris, The mechanism of the dendritic electro-crystallization of zinc, *J. Electrochem. Soc.*, 116 (1969), 1503–1514.
11. D.T. Chin, R. Sethi, J. McBreen, Zinc electrode morphology in alkaline solutions: I. Study of alternating voltage modulation on a rotating disk electrode, *J. Electrochem. Soc.*, 129 (1982), 2677–2685.
12. R.P. Bonner, Advanced primary zinc-air batteries, *Proceedings of the 23rd Annual Power Sources Conference*, Atlantic City, NJ, (May 1969).
13. G. Caprioglio, A. Weinberg, Materials science aspects of zinc–air batteries: A review, *Proceedings of the Sixth ZECEC Meeting*, Boston, MA, SAE, Warrendale, PA, (August 1971).
14. K. Wang, P. Pei, Z. Ma, H. Xu, P. Li, X. Wang, Morphology control of zinc regeneration for zinc-air fuel cell and battery, *J. Power Sources*, 271 (2014), 65–75.
15. S. Arouete, K.F. Blurton, H.G. Oswin, Controlled current deposition of zinc from alkaline solution, *J. Electrochem. Soc.*, 116 (1969), 166–169.
16. A.R. Despic, K.I. Popov, The effect of pulsating potential on the morphology of metal deposits obtained by mass-transport controlled electrodeposition, *J. Appl. Electrochem.*, 1 (1971), 275–278.
17. (a) K. Boto, Organic additives in zinc electroplating, *Electrodepos. Surface Treat.*, 3 (1975), 77–95. (b) B.C. Tripathy, S.C. Das, G.T. Hefter, P. Singh, Zinc electrowinning from acidic sulfate solutions. Part I: Effects of sodium lauryl sulfate, *J. Appl. Electrochem.*, 27 (1997), 673–678. (c) P.K. Leung, C. Ponce-de-León, C.T.J. Low, F.C. Walsh, Zinc deposition and dissolution in methanesulfonic acid onto a carbon composite electrode as the negative electrode reactions in a hybrid redox flow battery, *Electrochim. Acta*, 56 (2011), 6536–6546. (d) B.K. Thomas, D.J. Fray, The effect of additives on the morphology of zinc electrodeposited from a zinc chloride electrolyte at high current densities, *J. Appl. Electrochem.*, 11 (1981), 677–683. (e) J. Kan, H. Xue, S. Mu, Effect of inhibitors on Zn-dendrite formation for zinc-polyaniline secondary battery, *J. Power Sources*, 74 (1998), 113–116. (f) J.Y. Lee, J.W. Kim, M.K. Lee, H.J. Shin, H.T. Kim, S.M. Park, Effects of organic additives on initial stages of zinc

electroplating on iron, *J. Electrochem. Soc.*, 151 (2004), C25–C31. (g) O. Aaboubi, J. Douglade, X. Abenaqui, R. Boumedmed, J. VonHoff, Influence of tartaric acid on zinc electrodeposition from sulphate bath, *Electrochim. Acta*, 56 (2011), 7885–7889. (h) S.J. Banik, R. Akolkar, Suppressing dendrite growth during zinc electrodeposition by PEG-200 additive, *J. Electrochem. Soc.*, 160 (2013), D519–D523.

18. S.J. Banik, R. Akolkar, Suppressing dendritic growth during alkaline zinc electrode-position using polyethylenimine additive, *Electrochimica Acta*, 179 (2015), 475–481.

19. K. Miyazaki, Y.S. Lee, T. Fukutsuka, T. Abe, Suppression of dendrite formation of zinc electrodes by the modification of anion-exchange ionomer, *Electrochem.*, 80 (2012), 725–727.

20. J.F. Parker, C.N. Chervin, E.S. Nelson, D.R. Rolison, J.W. Long, Wiring zinc in three dimensions re-writes battery performance-dendrite-free cycling, *Energy Environ. Sci.*, 7 (2014), 1117–1124.

21. K. Kinoshita, *Electrochemical Oxygen Technology*, John Wiley & Sons, Inc., New York, (1992).

22. L. Jöerissen, Secondary batteries – metal-air systems: Bifunctional oxygen elec-trodes, *Encyclopedia of Electrochemical Power Sources*, (2009), 356–371. doi:10.1016/B978-044452745-5.00920-5.

23. (a) V. Neburchilov, H. Wang, J.J. Martin, W. Qu, A review on air cathodes for zinc-air fuel cells, *J. Power Sources*, 195 (2010), 1271–1291. (b) P. Xu, W. Chen, Q. Wang, T. Zhu, M. Wu, J. Qiao, Z. Chen, J. Zhan, Effects of transition metal precursors (Co, Fe, Cu, Mn, or Ni) on pyrolyzed carbon supported metal-aminopyrine electro-catalysts for oxygen reduction reaction, *RSC Adv.*, 5 (2015), 6195–6206.

24. (a) M.O. Davies, M. Clark, E. Yeager, F. Hovorka, The oxygen electrode: I. Isotopic investigation of electrode mechanisms, *J. Electrochem. Soc.*, 106 (1959), 56–61. (b) J. Shi, X. Zhou, P. Xu, J. Qiao, Z. Chen, Y. Liu, Nitrogen and sulfur co-doped mesoporous carbon material as highly efficient electrocatalysts for oxygen reduction reaction, *Electrochim. Acta*, 145 (2014), 259–269. (c) H.W. Park, D.U. Lee, P. Zamani, M.H. Seo, L.F. Nazar, Z. Chen, Electrospun porous nanorod perovskite oxide/nitrogen-doped graphene composite as a bi-functional catalyst for metal air batteries, *Nano Energy*, 10 (2014), 192–200.

25. (a) X. Wang, P.J. Sebastian, M.A. Smit, H. Yang, S.A. Gamboa, Studies on the oxy-gen reduction catalyst for zinc-air battery electrode, *J. Power Sources*, 124 (2003), 278–284. (b) D.U. Lee, H.W. Park, M.G. Park, V. Ismayilov, Z. Chen, Synergistic bifun-tional catalyst design based on perovskite oxide nanoparticles and intertwined carbon nanotubes for rechargeable zinc-air battery applications, *ACS Appl. Mater. Interfaces*, 7 (2014), 902–910.

26. R. Thacker, Some effects resulting from the use of a platinum catalyst in a zinc-oxygen cell, *Electrochim. Acta*, 14 (1969), 433–436.

27. (a) A.C.C. Tseung, S. Jasem, Oxygen evolution on semiconducting oxides, *Electrochim. Acta*, 22 (1977), 31–34. (b) W.J. King, A.C.C. Tseung, The reduction of oxygen on nickel-cobalt oxides – I: The influence of composition and preparation method on the activity of nickel-cobalt oxides, *Electrochim. Acta*, 19 (1974), 485–491.

28. F.W. Thomas Goh, Z. Liu, T.S.A. Hor, J. Zhang, X. Ge, Y. Zong, A. Yu, W. Khoo, A near-neutral chloride electrolyte for electrically rechargeable zinc-air batteries, *J. Electrochem. Soc.*, 161 (2014), A2080–A2086.

29. (a) M.F.M. Masri, A.A. Mohamad, Sago gel polymer electrolyte for zinc-air battery, *Adv. Sci. Tech.*, 72 (2010), 305–308. (b) R. Othman, W.J. Basirun, A.H. Yahaya, A.K. Arof, Hydroponic gel as a new electrolyte gelling agent for alkaline zinc-air cells, *J. Power Sources*, 103 (2001), 34–41. (c) J. Fu, J. Qiao, H. Lv, J. Ma, X.-Z. Yuan, H. Wang, Alkali doped poly(vinyl alcohol) (PVA) for anion-exchange membrane fuel cells: Ionic conduc-tivity, chemical stability and FT-IR characterization, *ECS Trans.*, 25 (2010), 15–23.

30. J.R. Goldstein, B. Koretz, Ongoing tests of the electric fuel zinc-air battery for electric vehicles, *IEEE Transactions on Industrial Electronics*, 0-7803-4098-1/98, (1998).

31. L.G. Danczyk, M.C. Cheiky, M.C. Wehrey, Second generation zinc-air powered electric minivans, *SAE International Congress and Exposition*, Detroit, MI, February 24–28, (1992).

32. P.N. Ross, A novel zinc-air battery for electric vehicles. *Proceedings of the Tenth Annual Battery Conference on Applications and Advances*, 1995, Long Beach, CA, 1995, pp. 131–133.

2 Lithium–Air Batteries
Fundamentals and Applications

Nobuyuki Imanishi and Osamu Yamamoto

CONTENTS

2.1 INTRODUCTION

Electric vehicles (EVs) with rechargeable batteries are considered to reduce CO_2 emissions and the consumption of fossil fuels because the total energy conversion efficiency of batteries is higher than that of internal combustion (IC) engines. The average tank to wheel energy efficiency of the U.S. fleet is 12.6%, while the efficiency of battery to wheel is approximately 90% [1]. Many types of batteries such as improved lead–acid [2], sodium–sulfur [2,3], metal–air [2], lithium–polymer [4], nickel–metal hydride [5], and lithium–ion [6] have been developed over the last half century. The first EVs were developed at the end of the nineteenth century and employed lead–acid batteries. The lead–acid battery was developed in 1859 by Planté. Lead–acid batteries are generally inexpensive and have a high power density; therefore, they have been used most widely for starting, ignition, and lighting of vehicles. The calculated energy density of lead–acid batteries is 171 Wh kg^{-1} and 370 Wh L^{-1} and the packed energy density of the sealed batteries for EVs is as low as 35–60 Wh kg^{-1} and 94–108 Wh L^{-1} [2]. The driving range for a compact EV with 14.4 kWh batteries is 110 km for a full charge. In 1989, Panasonic Ltd. and Sanyo Ltd. of Japan commercialized an advanced rechargeable nickel–metal hydride battery. The energy densities of the packed metal hydride battery were 48–63 Wh kg^{-1} and 82–142 Wh L^{-1}. The Toyota RAV4 EV with 27 kWh nickel–metal hydride batteries could drive for 178.5 km, where the 451 kg battery was 28.5% of the total car weight. In 1990, Sony Ltd., Japan, commercialized lithium–ion batteries with an even higher energy density. These batteries consisted of a carbon anode, a nonaqueous electrolyte, and a LiCoO$_2$ cathode. The reversible cell reaction is as follows [7]:

$$LiCoO_2 + 2.7C = Li_{0.55}CoO_2 + 0.45LiC_6 \qquad (2.1)$$

The cell voltage during operation is 3.9 V and the calculated specific energy density is 361 Wh kg^{-1}. The specific calculated energy density of a gasoline engine for automotive applications is estimated to be around 1700 Wh kg^{-1}, determined using the reaction heat of gasoline and a tank to wheel efficiency of 12.6% [1]. The calculated energy density of lithium–ion batteries is thus only one-fifth that of gasoline engines. To achieve EVs with comparable driving ranges to gasoline engine vehicles, battery systems with theoretical energy densities greater than 1700 Wh kg^{-1} should be developed. Lithium–ion batteries have been extensively used for the recent EVs because they have the highest energy density of batteries developed to date. However, the driving range of these EVs is still too short compared to vehicles with IC engines. The Nissan LEAF EV commercialized by Nissan Motors Ltd. was announced with driving range at full charge of 228 km without air conditioning and 130 km with it, where the total car weight was 1520 kg and the battery weight was around 240 kg. The specific energy density of the lithium–ion batteries used for the Nissan LEAF is around 100 Wh kg^{-1}, which is approximately 30% of the calculated specific energy density for the lithium–ion battery.

Table 2.1 shows the calculated energy densities for various types of battery systems. Many requirements, such as cost, safety, and materials resources, as well as the specific energy density, should be overcome for EV batteries. The highest energy density battery system consists of a lithium anode and fluorine cathode. However, such a practical cell has not yet been proposed. The second highest specific energy density is calculated for an aluminum and oxygen couple. The energy density of the primary aluminum battery for EVs was reported to be as high as 728 Wh kg^{-1}, and the driving range was recorded as 400 km with 204 kWh batteries [8]. However, a practical rechargeable aluminum–oxygen (or air) cell has not yet been proposed. The magnesium–oxygen system has the third highest theoretical energy density at 3991 Wh kg^{-1}, but no rechargeable magnesium–oxygen cell has been proposed. The magnesium anode with a high capacity of 2205 mAh g^{-1} is, however, an attractive anode. Aurbach et al. [9] reported a rechargeable magnesium

TABLE 2.1

Calculated Energy Density for Various Electrode Materials

| Anode | Cathode | OCV (V) | Energy Density | | Remark (Energy Density of Packed Cell, Wh kg^{-1}) |
			Wh kg^{-1}	Wh L^{-1}	
Li	F_2	6.10	6302	16,600	No report
Al	O_2	2.71	4116	14,100	Primary cell (728)
Mg	O_2	3.1	3991	12,200	No report
Li	O_2	2.96	3458	6,102	Secondary cell
Na	O_2	2.83	2680	4,484	Secondary cell
Li	S	2.8–2.4	2602	2,887	Secondary cell (~250)
Zn	O_2	1.6	1054	5,960	Secondary cell
Li	Cr_3O_8	4.0	889		Secondary cell
C	$LiCoO_2$	3.92	363	1,330	Secondary cell (100–200)

battery in 2000, where the cell consisted of a magnesium anode, an electrolyte of $Mg(AlCl_2BuEt)_2$ in tetrahydrofuran (THF), and a $Mg_xMo_2S_4$ cathode. Since this report, many research groups have studied the magnesium battery system, but no high-capacity cathode has been reported. The fourth highest energy density battery system is lithium–oxygen. This system is the most promising battery system with a high specific energy density, as shown in this chapter. The other system with an energy density over 1700 Wh kg^{-1} is the lithium–sulfur system, which is described fully in Chapter 5 of this book.

2.2 HISTORY

The concept of primary lithium–air batteries was proposed by Littaner and Tsai [10,11] in 1976 and 1977, where an alkaline aqueous solution was used as the electrolyte. Lithium metal reacts with water and the *in situ* formation of a surface oxide film on lithium metal retards rapid corrosion. Lithium–air batteries are capable of achieving a high specific power density; a current density of 200 mA cm^{-2} was achieved at 2.0 V [10]. However, at an open circuit potential (OCV) of 2.9–3.0 V, corrosion is rapid, which results in low energy conversion efficiency, and the cell cannot be electrically recharged. The rechargeable lithium–air cell was reported by Semkow and Sammells [12] in 1987. A fuel electrode of solid oxide fuel cells with a stabilized zirconia oxide ion conductor was substituted for the lithium electrode. The cell configuration is

$$Li_xFeSi_2/LiF–LiCl–Li_2O/\text{stabilized } ZrO_2/La_{0.89}Sr_{0.1}MnO_3/Pt, \text{ air} \qquad (I)$$

The lithium alloy electrode was immersed in lithium conducting molten salt contained in a stabilized zirconia tube coated with a perovskite-type $La_{0.89}Sr_{0.1}MnO_3$ oxide electrode. The cell was operated at 650°C–850°C. The unlithiated $FeSi_2$ anode was charged up to $Li_{10}FeSi_2$ at 20 mA cm^{-2} (based on the anode surface area) at

650°C. The OCV was ca. 2.4 V at 650°C. The theoretical energy density of the cell based on the following cell reaction,

$$Li_{10}FeSi_2 + 5O_2 = 5Li_2O_2 + FeSi_2, \tag{2.2}$$

is 1884 Wh kg^{-1} for the discharged state. A cell voltage of 1.2 V at 200 mA cm^{-2} during discharge at 800°C was reported. A similar concept was reported by Batalov and Arkhigov [13] in 1988 for a cell configuration of

$$Li/LiBe_2O_3-LiCl-Li_2CO_3/Li_{0.1}NiO \text{ (or } La_{0.7}Sr_{0.3}MnO_3/O_2) \tag{II}$$

$LiBe_2O_3-LiCl-Li_2CO_3$ is a molten lithium conducting electrolyte with a $LiBe_2O_3$ matrix. The cell was operated at 580°C–600°C and showed a cell voltage of 2 V at 100 mA cm^{-2}. The molten salt–type lithium–air (oxygen) cell could operate at high current density, but the long-term cycling performance of this type of cell has not been reported.

In 1996, Abraham and Jang [14] reported the first rechargeable lithium–oxygen battery at room temperature. They used a gel-type lithium ion conducting polymer electrolyte of $LiPF_6$ in polyacrylonitrile (PAN) with ethylene carbonate (EC) and propylene carbonate (PC). The gel-type electrolyte was developed as an electrolyte for lithium–ion batteries by Abraham and Alomgir [15]. The electrical conductivity for the gel-type electrolyte of ca. 10^{-3} S cm^{-1} at room temperature is comparable to that for nonaqueous lithium ion conducting solutions with a separator. The OCV of the lithium–oxygen cell was approximately 3 V. Analysis of the discharged carbon electrode using Raman spectroscopy suggested that the reaction product is mainly Li_2O_2. The charge and discharge curves for the cell (III),

$$Li/8 \text{ wt% } LiPF_6\text{-}12 \text{ wt% PAN-40 wt% PC-40 wt% EC/acetylene black/O}_2, \tag{III}$$

are shown in Figure 2.1. Cobalt phthalocyanine catalyst was used for the oxygen reduction reaction (ORR) and oxygen evolution reaction (OER). The cell was first discharged to a 1.5 V cutoff at 0.1 mA cm^{-2} to yield a capacity of 580 mAh g^{-1} for the 250 μm thick carbon electrode. In the subsequent charge to 4.0 V at 0.05 mA cm^{-2}, a capacity of 630 mAh g^{-1} was obtained. The specific discharge capacity was dependent on the thickness of the carbon electrode. The reaction product is deposited on the carbon electrode and the capacity is dependent on the surface area of the carbon electrode. After the first report by Abraham and Jang [14], only a few papers on lithium–air batteries were reported over the next 10 years. Read [16,17] studied the discharge capacity of the ORR on a carbon electrode for various types of nonaqueous electrolytes, including carbonate- and ether-based electrolytes. The high capacities for the ORR of 1633 mAh g^{-1} at 0.5 mA cm^{-2} and 2494 mAh g^{-1} at 0.05 mA cm^{-2} were observed for a Super P carbon black in an electrolyte of LiBr in 1,3 dioxolane-1,2 dimethoxyethane, which is several times higher than that of the cathode in lithium–ion batteries. However, the cycling performance was quite poor.

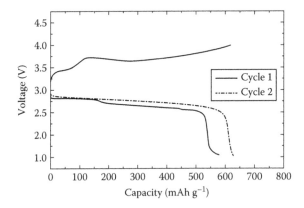

FIGURE 2.1 The cycling performance for Li/LiPF$_6$–PAN–EC–PC/acetylene black with phthalocyanine (250 μm thick), O$_2$ at room temperature. The cell was discharged at 0.1 mA cm^{-2} with a cutoff voltage of 1.0 V and charged at 0.05 mA cm^{-2} with a cutoff voltage of 4.0 V. (From Abraham, K.M. and Alomgir, M., *J. Electrochem. Soc.*, 137, 1657, 1990.)

In 2006, Bruce and coworkers [18] reported a rechargeable nonaqueous lithium–oxygen cell with an electrolytic manganese oxide (EMD) catalyst for the oxygen electrode, after which the research activity on rechargeable lithium–air batteries significantly increased. Figure 2.2 shows the first cycle charge and discharge performance and discharge capacity change with cycles for cell (IV):

$$\text{Li/LiPF}_6 \text{ in PC/Super P carbon-EMD/O}_2 \qquad (IV)$$

Bruce and coworkers [18] concluded that the electrochemical reaction of $2\text{Li}^+ + 2\text{e}^- + \text{O}_2 = \text{Li}_2\text{O}_2$ in a nonaqueous lithium battery with an O$_2$ cathode is reversible and the charge/discharge cycling can be sustained for over tens of cycles. The formation of Li$_2$O$_2$ was confirmed using *in situ* differential electrochemical mass spectrometry (DEMS). However, these results suggest many problems, such as the mechanism of capacity degradation by cycling and the high polarization for the ORR and OER. These problems will be discussed in later sections.

2.3 TYPE OF LITHIUM–AIR BATTERIES

Currently, two types of rechargeable lithium–air systems, nonaqueous and aqueous, are under development. Schematic diagrams of the cell configurations are shown in Figure 2.3. Most research on lithium–air (oxygen) batteries has been focused on the nonaqueous system with an organic aprotic electrolyte solution. The nonaqueous system consists of a lithium anode, a nonaqueous electrolyte, and an air electrode. The cell reaction for the nonaqueous system is considered to be the formation of Li$_2$O$_2$ by the combination of reduced oxygen with lithium ion from the electrolyte [15,18], and the process is reversed on charging:

$$2\text{Li} + \text{O}_2 = \text{Li}_2\text{O}_2 \qquad (2.3)$$

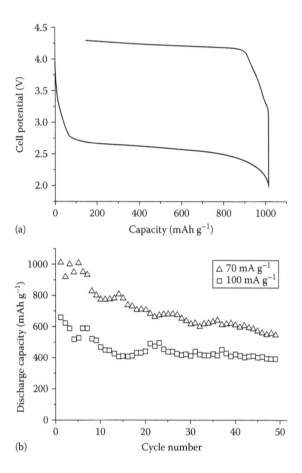

FIGURE 2.2   (a) Charge and discharge curves for a Li/PC–LiPF$_6$/Super P carbon-EMD-PVDF,O$_2$ cell at 50 mA g^{-1}. (b) Variation of discharge capacity with cycle number; rate=70 and 100 mA g^{-1}. Capacities are expressed per gram of carbon in the electrode. (From Ogasawara, T. et al., *J. Am. Chem. Soc.*, 128, 1390, 2006.)

The reversible cell voltage calculated from the standard Gibbs free energy is 2.96 V [19]. The calculated mass specific energy density using the weight of lithium and oxygen and the OCV is 3458 Wh kg^{-1}, and the calculated volume specific energy density using the volume of lithium metal is 6102 Wh L^{-1}. The calculated mass energy density is around one order higher than that for conventional lithium–ion batteries. Aprotic solvents with a dissolved lithium salt have been used widely as nonaqueous electrolytes. Ionic liquids are also an attractive candidate for the electrolyte in nonaqueous lithium–air batteries. These electrolytes will be introduced in a later section.

Solid electrolytes have also been proposed for the electrolyte in lithium–oxygen cells. Kumar et al. [20,21] have used a NASICON-type lithium ion

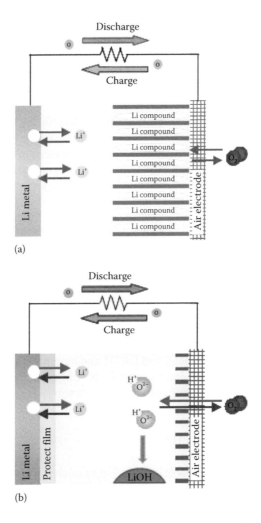

FIGURE 2.3 Schematic cell configurations for (a) nonaqueous and (b) aqueous lithium–air cells.

conducting solid electrolyte of lithium aluminum germanium phosphate glass ceramic, $Li_{1+x}Al_xGe_{2-x}(PO_4)_3$ (LAGP), the electrical conductivity of which was reported to be as high as 4.48×10^{-3} S cm^{-1} at 23°C [22]. However, this type of solid lithium conductor is unstable in contact with lithium metal, and a polyethylene oxide–based polymer electrolyte was used as an interlayer between the lithium metal and LAGP to prevent direct contact of these components. However, the lithium ion conductivity of the polymer electrolyte is low at room temperature; therefore, the cell was operated at a higher temperature. An areal specific capacity of 5.7 mAh cm^{-2} was observed at 0.1 mA cm^{-2} and 95°C. The charge and discharge columbic efficiency at the first cycle was less than 70%, and the cycling performance of the cell was not presented. Scrosati and coworkers [23] proposed a polymer electrolyte lithium–oxygen cell with a composite polymer electrolyte of poly(ethylene oxide) (PEO)–$LiCF_3SO_3$-nano-ZrO_2. The Li/polymer electrolyte/Super P carbon black, O_2 cell showed an OCV of 3.2 V and the discharge product was confirmed to be Li_2O_2 by X-ray diffraction (XRD) analysis. However, the detailed charge and discharge performance were not reported.

In the aqueous system proposed by Littaner and Tsai [10], the lithium metal electrode was in direct contact with the aqueous electrolyte, where lithium reacts with water. The anode oxide film formed on the lithium surface permits an inert cathode structure. However, hydrogen evolution at the lithium surface always creates pores within the film. Thus, the current efficiency of this type aqueous lithium–air cell is governed by the ratio of two competing reactions: the ORR at the cathode and the corrosion reaction at the anode. A new concept of rechargeable aqueous lithium–air batteries was proposed by Visco et al. in 2004 [24]. A schematic diagram of the aqueous lithium–air battery is shown in Figure 2.3b. The lithium metal electrode was protected by a water-stable NASICON-type lithium ion conducting solid electrolyte. The high lithium ion conductivity of the NASICON-type solid electrolyte of $Li_{1+x}Al_xTi_{2-x}(PO_4)_3$ (LATP) was reported by Aono et al. [25] in 1990. The highest electrical conductivity of this system was reported to be 7×10^{-4} S cm^{-1} at 25°C. However, LATP is unstable in contact with lithium metal. Therefore, an interlayer between LATP and lithium metal is required. The interlayer should be stable in contact with lithium metal and LATP, and have high lithium ion conductivity. Visco et al. proposed an interlayer of Li_3N, while Imanishi and colleagues [26] proposed a lithium conducting polymer electrolyte interlayer. The cell reaction for the aqueous system is

$$4Li + 6H_2O + O_2 = 4(LiOH \cdot H_2O) \tag{2.4}$$

The LiOH reaction product is soluble in water, but its solubility in water is 5.3 mol L^{-1}, which corresponds to a discharge depth of ca. 30%. LATP is unstable in saturated LiOH aqueous solution, but stable in saturated LiOH aqueous solution with saturated LiCl [27]. The OCV for the aqueous system with saturated LiOH and saturated LiCl is 2.98 V [28]. The specific mass and volume energy densities of the aqueous system calculated from reaction (2.4) and the OCV of 2.98 V are 1897 Wh kg^{-1} and 1991 Wh L^{-1}, respectively. These specific energy densities are lower than those of the nonaqueous system because water in the electrolyte is an active material for the cell reaction. However, the mass specific energy density of aqueous lithium–air batteries is approximately five times higher than that of conventional lithium–ion batteries. The calculated mass specific energy density of the aqueous system is comparable with that calculated for the IC engine. A report [1] distinguished the aqueous system and mixed aqueous aprotic system, but these two systems show the same cell reaction (2.4), and at present an aprotic electrolyte between lithium metal and the solid electrolyte should be used, because the acceptable lithium conducting solid electrolyte is unstable in contact with lithium metal. After a water- and lithium-stable lithium conducting solid electrolyte is developed, the aprotic electrolyte will not be used for the aqueous system.

2.4 NONAQUEOUS LITHIUM–AIR SYSTEM

2.4.1 ELECTROCHEMICAL REACTION IN THE NONAQUEOUS LITHIUM–AIR SYSTEM

The electrochemical electrode reactions of the nonaqueous system [14,18] are considered to be

$$Li = Li^+ + e \quad \text{(anode)} \ (E^\circ = -3.04 \text{ V vs. NHE}) \tag{2.5}$$

and

$$2Li + O_2 = Li_2O_2 \quad \text{(cathode)} \tag{2.6}$$

Reaction (2.6) is considered to proceed by several steps [29–31]:

$$O_2 + e^- = O_2^- \quad (E^\circ = -0.563 \text{ V vs. NHE}) \tag{2.7}$$

$$O_2^- + Li^+ = LiO_2 \tag{2.8}$$

$$LiO_2 + LiO_2 = Li_2O_2 + O_2 \tag{2.9}$$

This mechanism is the Li^+ analogue of a conventional proton-induced charge transfer in the aqueous case [30]. *In situ* surface-enhanced Raman spectroscopy has provided direct evidence that oxygen reduction on Au in oxygen saturated 0.1 M $LiClO_4$–CH_3CN [31]. The standard Li_2O_2 formation potential is 2.96 V versus Li^+/Li. Lu et al. [29] suggested that the formation of Li_2O during a deep discharge or with a reactive metal catalyst.

In 2010, Mizuno et al. [32] reported the decomposition of a carbonate-based liquid electrolyte of $Li(CF_3SO_2)_2N$ (LiTFSI) in PC using transmission electron microscopy (TEM) observation and Fourier transform infrared spectroscopy (FTIR) spectroscopy. Energy-dispersive x-ray spectroscopy analysis of the discharge product showed both signals of carbon and oxygen, which indicates that the formation of a carbonate. The FTIR spectra of the discharge products indicated the formation of Li_2CO_3 and lithium alkyl carbonates. However, peaks due to RO–(C=O)–OLi (R = alkyl group) and Li_2O_2 were not clearly observed. These results suggested electrolyte decomposition by the formation of superoxide ions and/or lithium superoxide radicals by Equations 2.7 and 2.8. In 2011, Bruce and coworkers [33] concluded that a nonaqueous rechargeable lithium–oxygen battery containing alkyl carbonate electrolytes was discharged by the formation of $C_3H_6(OCO_2Li)_2$, Li_2CO_3, HCO_3Li, CH_3CO_2Li, CO_2, and H_2O at the cathode due to electrolyte decomposition. Charging involved the oxidation of $C_3H_6(OCO_2Li)_2$, Li_2CO_3, HCO_3Li, and CH_3CO_2Li. To detect the reaction products, they used many techniques, such as *in situ* DEMS [34], *in situ* surface enhanced Raman spectroscopy (SERS), and FTIR spectroscopy. The different pathways for charge and discharge are consistent with the widely observed voltage gap in lithium–oxygen cells. The proposed reaction schemes for discharge and charge are shown in Figures 2.4 and 2.5, respectively. Luntz and colleagues [35]

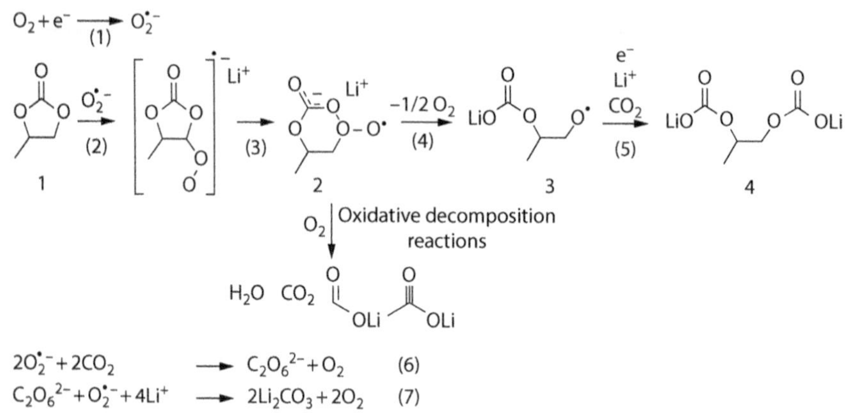

$$O_2 + e^- \xrightarrow{(1)} O_2^{\bullet-}$$

$$2O_2^{\bullet-} + 2CO_2 \longrightarrow C_2O_6^{2-} + O_2 \quad (6)$$
$$C_2O_6^{2-} + O_2^{\bullet-} + 4Li^+ \longrightarrow 2Li_2CO_3 + 2O_2 \quad (7)$$

FIGURE 2.4 Proposed reaction scheme for discharge of the lithium–oxygen cell with 1 M LiPF$_6$ in PC. (From Freunberger, S.A. et al., *J. Am. Chem. Soc.*, 133, 8040, 2011.)

also studied the lithium–oxygen electrochemistry in various solvents, including carbonate (typical lithium–ion battery solvent) and 1,2-dimethoxyethane (DME) with 1 M Li(CF$_3$SO$_2$)$_2$N (LiTFSI) using DEMS, coupled with isotopic labeling of oxygen gas. They found that the carbonate-based solvents were formed by irreversible decomposition upon cell discharge. However, the cell with the DME organic ether solvent produced mainly lithium superoxide upon discharge. Upon cell charge, lithium superoxide both decomposed to evolve oxygen and oxidize DME at high potentials. Comparing the OER during charge to the oxygen consumption during discharge (ORR) showed that the electrochemical reversibility (OER/ORR) was only ca. 80% for DME. From these results, it could be concluded that carbonate solvents as electrolytes in lithium–oxygen batteries, which were widely used as the solvent in the electrolyte during the early stage of the rechargeable lithium–air battery research, are unstable for the superoxide produced during the discharge process and not suitable for use in rechargeable lithium–air batteries. Ether-based solvents such as DME have been reported to be stable with the Li$_2$O$_2$ peroxide but unstable with the LiO$_2$ lithium superoxide. After these reports, many other types of solvents have been proposed for lithium–oxygen batteries. In later sections, the status and possibility of a stable nonaqueous electrolyte for lithium–oxygen batteries will be introduced.

2.4.2 STABILITY OF NONAQUEOUS ELECTROLYTES FOR RECHARGEABLE LITHIUM–AIR BATTERIES

In the rechargeable lithium–oxygen battery, the lithium superoxide (LiO$_2$) is a primary intermediate, as shown in reaction (2.8), which was confirmed using *in situ* SERS by Bruce and coworkers [31] for 0.1 M *n*-BuNClO$_4$ in CH$_3$CN. LiO$_2$ could be either converted to Li$_2$O$_2$ or involved in decomposition of the electrolyte. Bryantsev et al. [36,37] reported the free energy barriers (ΔG_{act}) and reaction free

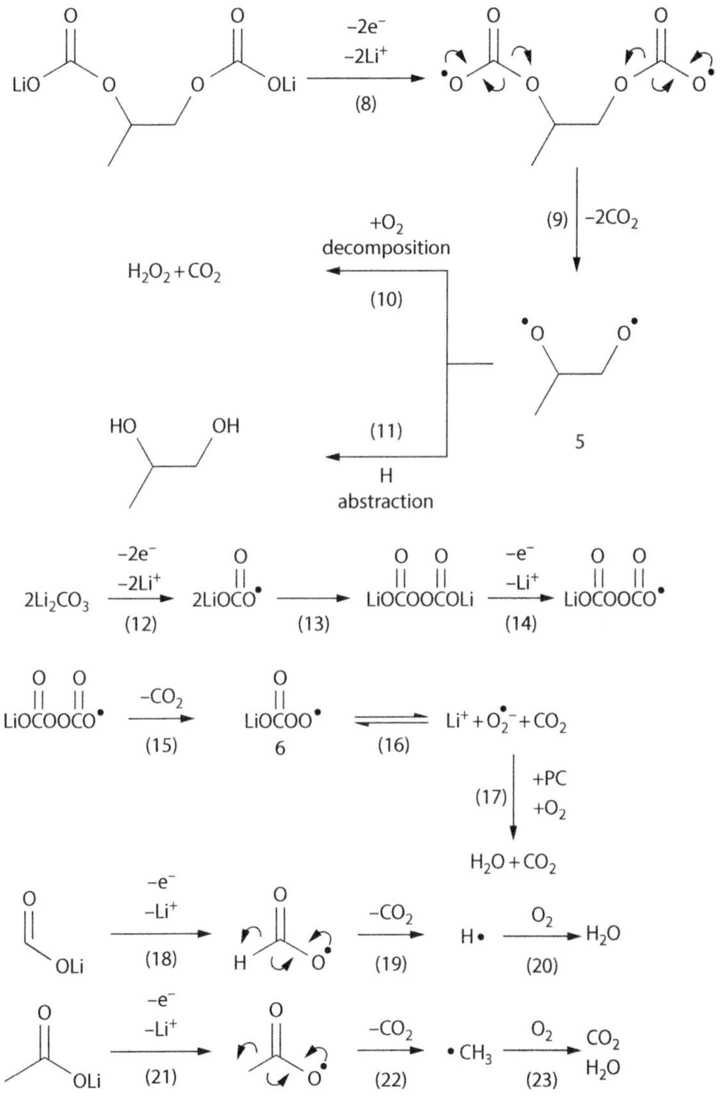

FIGURE 2.5 Proposed reaction scheme for charge of the lithium–oxygen cell with 1 M LiPF$_6$ in PC. (From Freunberger, S.A. et al., *J. Am. Chem. Soc.*, 133, 8040, 2011.)

energy (ΔG_r) calculated for the nucleophilic substitution of a series of aprotic solvents by the superoxide anion. These data provide a convenient means of screening for solvents that are suitable for lithium–air batteries from first principles calculations. Species that have chemical functionality and stability against nucleophilic substitution by superoxide ions include *N*-alkyl subsided amides, lactones, nitriles, and ethers. Calculations revealed the exceptional nucleophilic stability of *N,N*-dialkyl amides and the electrochemical behavior of an O$_2$ cell in a dimethyl amide

(DMA)–based electrolyte with 0.5 M LiTFSI composed of a Super P carbon black electrode and a LiFePO$_4$ counter electrode. The results showed the repeated formation of lithium peroxide upon discharge and its reversible decomposition upon charge. The stability of the electrolyte was confirmed by *in situ* gas analysis of the lithium–O$_2$ cell. However, DMA is unstable in contact with lithium metal and the cyclic performance of the Li/LiTFSI in DMA/LiFePO$_4$ cell is poor. Bruce and colleagues [38] also reported the stability of dimethylformamide (DMF) as an electrolyte. Reactions at the O$_2$ cathode during the first discharge–charge cycle were dominated by reversible Li$_2$O$_2$ formation and decomposition, but upon further cycling, electrolyte decomposition was observed using FTIR and *in situ* DEMS, as shown in Figure 2.6.

In 2011, many research groups reported the possibility of electrolytes based on ethers [35,39–43]. Ethers are attractive for the electrolyte of lithium–air cells because they are stable in contact with lithium metal and stable up to an oxidation potential of 4 V. In 2006, Read [17] reported the discharge performance of a lithium–oxygen cell with an ether-based electrolyte composed of a mixture of 1,3-dioxolane (DOL) and DME with various lithium salts. The viscosity of the ether-based electrolyte is lower than that of the carbonate electrolytes, and a high discharge capacity of 1633 mAh g^{-1} was obtained at a high current density of 0.5 mA cm^{-2} using a Super P carbon oxygen electrode. However, the stability of the electrolyte for the superoxide was not reported. Luntz and coworkers [35] reported the stability of an ether-based electrolyte of LiTFSI in DME, EC-DME, and PC-DME using DEMS coupled with isotopic labeling of oxygen gas. Figure 2.7 shows the isotopically labeled O$_2$ and CO$_2$ gas evolution rate during charging of DME, EC/DME (1:1 v/v), and PC/DME (1:2 v/v). As expected, the DME-based electrolyte discharged under ^{18}O$_2$ predominately evolved ^{18}O$_2$ upon charging, which indicated that the O$_2$ evolved was derived exclusively from O$_2$ consumed by the cell during discharge. The total O$_2$ evolved in the DME-based cell corresponded to only 60% evolution of that expected from 1 mAh discharge.

A possible explanation for the low O$_2$ evolution during charge could be a slow thermal, chemical, or electrochemical reaction between DME and Li$_2$O$_2$ that consumes a portion of the Li$_2$O$_2$ prior to or during cell discharge. The CO$_2$ isotopes evolved at the end of the charge were in a 45:40:15 molar ratio for C^{18}O$_2$:C^{16}O^{18}O:C^{16}O$_2$. The large molar concentration of ^{18}O present in the CO$_2$ isotope indicates that Li$_2$^{18}O$_2$ formed during discharge participates in DME oxidation at high voltage (4.5–4.6 V) and that under these conditions the electrochemical stability of the electrolyte in the presence of Li$_2$O$_2$ is poor. For the EC/DME (1:1 v/v) and PC/DME (1:2 v/v) cells, CO$_2$ was by far the dominant species formed during charging; the molar ratio of total O$_2$:CO$_2$ generated during charge was 1:27 for the EC/DME (1:1 v/v)-based electrolyte and 1:9 for the PC/DME (1:2 v/v)-based electrolyte. The isotopic composition of the CO$_2$ evolved was 5:70:25 for C^{18}O$_2$:C^{16}O^{18}O:C^{16}O$_2$. These results clearly indicate that electrolyte decomposition occurs because the only available source of ^{16}O is the electrolyte. Consequently, carbonate-based electrolytes undergo chemical and electrochemical reduction in the presence of Li$_2$O$_2$ or the LiO$_2$ superoxide. In contrast, the DME-based electrolyte produces the appropriate rechargeable discharge product of Li$_2$O$_2$; however, DME is oxidatively unstable during recharge in the presence

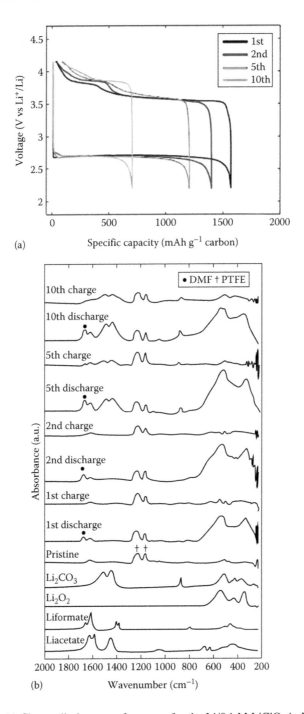

(a)

(b)

FIGURE 2.6 (a) Charge discharge performance for the Li/0.1 M LiClO₄ in DMF/Super P carbon-PTFE/O₂ cell and (b) FTIR spectra of Super P carbon-PTFE composite electrode cycled in 0.1 M LiClO₄ in DME. (From Chen, Y. et al., *J. Am. Chem. Soc.*, 134, 7952, 2012.)

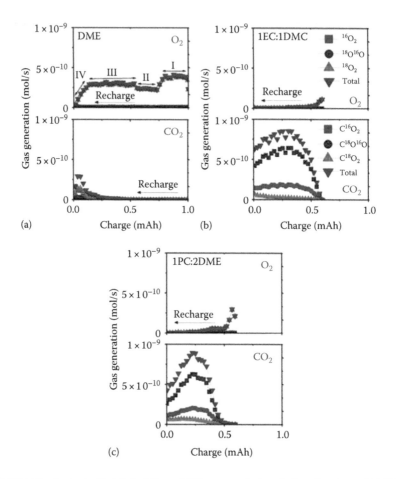

FIGURE 2.7 Isotopically labeled O_2 and CO_2 gas evolution during charging of (a) DME-based, (b) 1:1 (v/v) EC/DME-based, and (c) 1:1 (v/v) PC/DME-based Li–O_2 cells. (From McCloskey, B.D. et al., *J. Phys. Chem. Lett.*, 2, 1161, 2011.)

of Li_2O_2. Luntz and colleagues [35] concluded that both DME and carbonates are unstable solvent for secondary lithium–air batteries.

An electrolyte with low vapor pressure is useful to enable the operation of lithium–air batteries without solvent loss when O_2 is supplied into the cell from the open air. Tetraethylene glycol dimethyl ether $CH_3O(CH_2CH_2O)_4CH_3$ (TEGDME) is a such a candidate solvent for the electrolyte of lithium–air batteries. A lithium–O_2 cell consisting of a lithium metal anode, an electrolyte of 1 M $LiPF_6$ in TEGDME, and a porous Super P carbon black cathode was discharged to 2 V and charged to 4.6 V at 70 mA g^{-1} in 1 atm of O_2 [39]. Figure 2.8 shows the XRD patterns for the oxygen electrode cycled between 2 and 4.6 V versus Li/Li⁺, and the charge–discharge curves. The XRD data indicated the presence of Li_2O_2 at the end of the first discharge. However, the FTIR spectra provide clear evidence of electrolyte decomposition, although it is difficult to identify all the specific discharge products by FTIR alone. ¹H nuclear

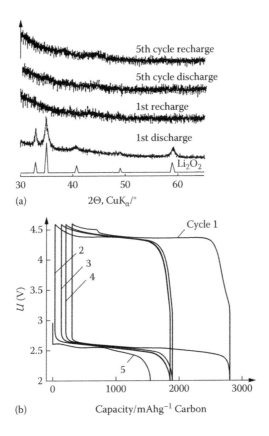

FIGURE 2.8 (a) XRD patterns of the composite cathode cycled in 1 M LiPF$_6$ in TEGDME under 1 atm O$_2$ between 2 and 4. 6 V versus Li/Li$^+$ at 70 mA g^{-1} and (b) load curves for the same cell. (From Freunberger, S.A. et al., *Angew. Chem. Int. Ed.*, 50, 8609, 2011.)

magnetic resonance spectroscopy (NMR) analysis of the discharge products washed with D$_2$O suggested the presence of HCOOD, and CH$_3$COOD. Mass spectroscopy analysis showed that CO$_2$ and H$_2$O were evolved.

Cyclic ethers, such as DOL and 2-methyl-THF, were in turn used to replace TEGDME in lithium–oxygen cells [39]. The XRD data for the discharge products of the oxygen electrode with these electrolytes indicated the formation of Li$_2$O$_2$, and the FTIR and NMR data demonstrated the presence of HCOOLi and CH$_3$COOLi. In conclusion, electrolytes based on linear and cyclic ethers all exhibit electrolyte decomposition. A comparison of ether-based electrolytes with organic carbonate-based electrolytes indicates that the ether-based electrolytes are more stable toward reduced O$_2$ species than the organic carbonate electrolytes. Abraham and colleagues [40] also examined the charge and discharge performance of a Li/LiPF$_6$ in TEGDME/carbon black, O$_2$ cell. The XRD pattern of the carbon electrode after discharge to 2.0 V indicated the presence of Li$_2$O$_2$; however, the rechargability at 0.13 mA cm^{-2} was poor.

Dimethyl sulfoxide (CH$_3$SOCH$_3$; DMSO) is the other candidate solvent for use as a nonaqueous electrolyte for lithium–air batteries [44], which has a very high

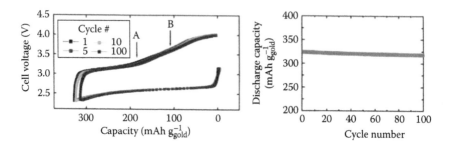

FIGURE 2.9 Charge/discharge curves and cycle profile for a Li–O$_2$ cell with a 0.1 M LiClO$_4$-DMSO electrolyte and a NPG cathode at a current density of 500 mA g^{-1} (based on mass of Au). (From Peng, Z. et al., *Science,* 337, 563, 2012.)

Gutmann donor number. In 2012, Bruce and coworkers [45] reported excellent cyclic performance for a lithium–air cell composed of a DMSO-based electrolyte and a nanoporous gold (NPG) electrode, where NPG was used for the oxygen electrode to avoid the effect of carbon decomposition, because carbon is unstable when charging above 3.5 V [46]. The cyclic performance for the Li/0.1 M LiClO$_4$ in DMSO/NPG/O$_2$ cell is shown in Figure 2.9. The initial discharge capacity of 95% was retained after 100 cycles. The oxygen electrode after cycling was analyzed using FTIR, SERS, and DEMS. Small amounts of Li$_2$CO$_3$ and HCOOLi were observed in the FTIR spectra at the end of discharge. The proportion of these reaction products to Li$_2$O$_2$ was estimated to be less than 1%. The DEMS analysis showed that the detected gas on discharge was O$_2$. The lack of evidence for CO$_2$ and SO$_2$ suggests that there was no decomposition of the electrolyte in contrast to the other electrolytes. The O$_2$ consumption during discharge was measured using DEMS and was almost the same as that calculated from the discharge capacity. The cycling performance for a Li/LiPF$_6$ in DMSO/porous carbon black (Ketjen black; KB), O$_2$ cell exhibited capacity fading with cycling [47]. The degradation of cell performance by cycling could be explained by the effect of carbon because carbon promotes electrolyte decomposition during charge and is unstable on charge [45,46]. In 2014, Shao-Horn and coworkers [48] reported the stability of DMSO in lithium–air batteries. The test cell consisted of a lithium metal anode, 0.1 M LiClO$_4$ in DMSO as an electrolyte, and a carbon nanotube O$_2$ electrode. Li$_2$O$_2$ was the only species detected by XRD analysis immediately after discharge, but this was gradually decomposed into LiOH after prolonged exposure to the electrolyte. Commercial Li$_2$O$_2$ power could decompose DMSO and the presence of KO$_2$ accelerated DMSO decomposition to dimethyl sulfone (DMSO$_2$). While only Li$_2$O$_2$ was detected immediately after discharge, only LiOH was found after 300 h aging. These results allowed us to probe the chemical stability of DMSO with respect to the oxygen reduction product, without the inference of a carbon electrode. Shao-Horn and coworkers [48] concluded that DMSO may not be suitable for the oxygen electrode in the development of rechargeable lithium–air batteries with long cycle life. In addition, DMSO electrolyte may not be stable in contact with lithium metal [45].

Ionic liquids are also considered as candidates for the electrolyte in rechargeable lithium–air batteries. In 2005, Kuboki et al. [49] reported the discharge performance for a lithium–air cell with LiTFSI in an ionic liquid of 1-ethy-3-methylimidazolium bis(trifluoromethylsulfonyl)imide. The cell showed a steady cell voltage at 0.01 mA cm^{-1} for 56 days under an air atmosphere. This result suggests that ionic liquids could be stable for the superoxides. Passerini and colleagues [50] also reported high electrochemical stability for a lithium conducting ionic liquid of N-butyl-N-methylpyrrolidinium bis(trifluoromethylsulfonyl)imide-LiTFSI against superoxide formation using cyclic voltammetry. Rechargeable lithium–air batteries with several types of ionic liquids, such as pyrrolidinium-based [51,52] and piperidinium-based [53–55] ionic liquids, have recently been studied by some groups. A high reversible capacity of 2000 mAh at 0.1 mA cm^{-1} and 60°C was obtained for the N-methyl-N-propylpiperidinium-bis(trifluoromethylsulfonyl)imide-LiTFSI [54]. However, the stability of this ionic liquid with superoxide is questionable. Recently, Scrosati and colleagues [52] reported an excellent cyclic performance for the lithium–O$_2$ cell with the N-butyl-N-methylpyrrolidinium bis(trifluoromethylsulfonyl)imide (PYR)-LiTFSI electrolyte. Figure 2.10 shows charge and discharge profiles for the Li/PYR$_{18}$LiTFS/Super P carbon/O$_2$ cell at 0.08 mA cm^{-2} and 30°C. A high area specific capacity of 4 mAh cm^2 (4500 mAh g$_{carbon}^{-1}$) was obtained. Li 1s XPS spectra of the carbon electrode at the end of discharge to 1000 mAh g$_{carbon}^{-1}$ showed a signal for Li$_2$O$_2$, as shown in Figure 2.10b. The charge and discharge performance of the cell under a limited capacity of 500 mAh g$_{carbon}^{-1}$ showed no significant degradation for 30 cycles. However, details on the stability of the ionic liquid in the presence of the superoxide prepared at the carbon electrode have not been reported. However, Gasteiger and colleagues [51] claimed that pyrrolidinium ions are reduced on metallic lithium and produce a substantial amount of alkenes and amines and also may not have sufficient long-term stability against attack from the superoxide during discharge or from oxygen species produced during charge.

At present, no electrolyte that exhibits 100% stability for the side reactions in the ORR and OER has been reported. There are still serious problems to be solved for the nonaqueous electrolytes of lithium–air batteries. Such batteries for EV applications should have lifetimes of 10 years or more; therefore, a 500 cycle life would suffice. Therefore, research on electrolytes with focus on the degradation rate is the best course of direction, as pointed out by Bruce and colleagues [56].

2.4.3 CELL PERFORMANCE OF RECHARGEABLE NONAQUEOUS LITHIUM–AIR BATTERIES

The performance of lithium metal anodes in nonaqueous electrolytes has been studied in rechargeable lithium batteries since the 1970s [57]. However, there are several severe problems that include lithium dendrite formation and limited columbic efficiency during repeated lithium deposition and stripping processes. Recently, Zhang and coworkers [58] have reported that a highly concentrated composition of DME and Li(FSO$_2$)$_2$N (LiFSI) enables a high rate lithium metal anode at high columbic efficiency (up to 99.1%) without lithium dendrite growth. However, this ether-based

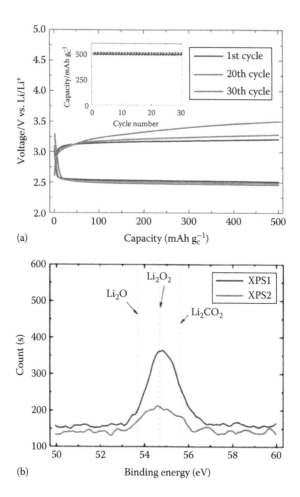

FIGURE 2.10   (a) Voltage profiles of the Li/PYR$_{14}$TFSI-LiTFSI/C, O$_2$ cell at 50 mA g^{-1} and 30°C, and (b) Li 1s spectra of the carbon electrode discharged to 1000 mAh g^{-1} (XPS1) and after charge (XPS2). (From Ella, G.A. et al., *Nano Lett.*, 14, 6572, 2014.)

electrolyte is unstable with the superoxide formed in the ORR at the air electrode. Therefore, nonaqueous electrolytes that suppress lithium dendrite growth and that are stable in contact with the superoxide should be developed.

The oxygen electrode reactions in aqueous electrolytes have been extensively studied since the 1970s [59]. There were very few reports on the oxygen electrode in nonaqueous electrolytes before the lithium–air cell, except for those of molten carbonates and solid oxide electrolytes at high temperature. In the earlier studies on nonaqueous lithium–air batteries, the carbonate electrolytes were mainly used and the cell performance included the decomposition reaction of the electrolyte. The decomposition products, such as Li$_2$CO$_2$, disturb further oxygen electrode reactions. As shown in Figure 2.2, the poor cyclability is due to the effect of electrolyte decomposition.

After 2011, almost all reports on lithium–oxygen cells have used more stable non-aqueous electrolytes, such as DME, TEGDME, and DMSO. Abraham and coworkers [40] reported the cell performance of cell (V):

$$\text{Li}/1\text{M LiPF}_6 \text{ in TEGDME/BP2000 carbon black, O}_2 \qquad\qquad (V)$$

The cell was discharged and charged for 14 h at 0.13 mA cm^{-2}. A high discharge and charge capacity was maintained during the first several cycles, but the cell capacity decreased with the increasing charge voltage polarization after four cycles. Bruce and colleagues [39] also examined the cyclic performance of a Li/1 M LiPF$_6$ in TEGDME/Super P carbon, O$_2$ cell and observed a significant capacity fade after few cycles as shown in Figure 2.8b. The capacity fade is most likely due to the accumulation of Li$_2$CO$_3$ on the carbon electrode surface, which is more difficult to oxidize. Li$_2$CO$_3$ is considered to be a final decomposition product of the electrolyte and carbon electrode [46]. Scrosati and colleagues [60] reported excellent cyclic performance for a lithium–oxygen cell using a TEGDME-LiCF$_3$SO$_3$ electrolyte, which exhibited high capacity and stable cyclic performance at a high current density. Figure 2.11 shows the cyclic performance of cell (VI):

$$\text{Li}/1 \text{ M LiCF}_3\text{SO}_3 \text{ in TEGDME/Super P carbon/O}_2 \qquad\qquad (VI)$$

The specific capacity of the cell is based on the mass of carbon, where the electrode packing density was 0.125 g cm^{-3}. This cell exhibited very stable cycling performance up to 100 cycles with a capacity of 1000 mAh g^{-1} at 1000 mA g^{-1} (1 mA cm^{-2}), and up to 300 cycles with a capacity of 5000 mAh g^{-1} at 50 mA g^{-1} (0.05 mA cm^{-2}). As shown in Section 2.4.2, FTIR spectra provided clear evidence of the decomposition of LiPF$_6$ in TEGDME, and mass spectroscopy analysis showed that CO$_2$ and H$_2$O were evolved from the cell during the charge and discharge processes. Amine and coworkers [61,62] provided evidence that the stability of the electrolyte in the lithium–O$_2$ cell is strongly dependent on the compatibility of the lithium salt with the solvent. Recently, Scrosati and colleagues [63] reported the effect of lithium salt in TEGDME for the cyclic performance of Li/1M LiX (X=PF$_6$, ClO$_4$, CF$_3$SO$_3$, and (CF$_3$SO$_2$)$_2$N in TEGDME/Super P carbon/O$_2$ cells at 200 mA g^{-1} and 25°C for a discharge capacity of 1000 mAh g^{-1}. No degradation of the charge and discharge performance was observed for 10 cycles, except for the cell with an electrolyte of LiPF$_6$ in TEGDME. However, the stability of LiX in TEGDME for the superoxide produced at the air electrode should be examined quantitatively using DEMS, as pointed out by Luntz et al. [64].

The other candidate for the electrolyte of reversible lithium–air batteries is DMSO, which was selected by Takechi et al. using a KO$_2$ solvent screening method [65] and a brief cyclic performance was reported for the Li/1 M LiTFSI in DMSO/C, O$_2$ cell. The reaction product of Li$_2$O$_2$ was observed using a Raman spectral analysis. Bruce and colleagues showed the stability of 0.1 M LiClO$_4$ in DMSO using a gold nanofiber oxygen electrode as shown in Figure 2.9 [45]. However, Abraham and coworkers [47] reported a poor cyclic performance for the Li/1 M LiPF$_6$ in DMSO/KB/O$_2$ cell at a discharge current density of 0.1 mA cm^{-2} and charge current density

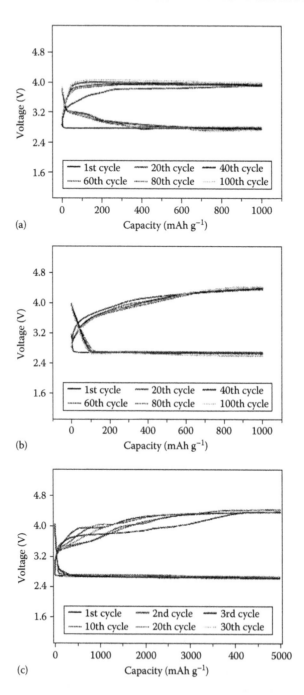

FIGURE 2.11 Cyclic performance of the Li/LiCF$_3$SO$_3$ in TEGDME/Super P carbon/ O$_2$ cell under various specific capacity limits and current densities of (a) 50, (b) 1000, and (c) 500 mA g^{-1}, where the specific capacity and current density based on the mass of the carbon. (From Jung, H.-G. et al., *Nat. Chem.*, 4, 579, 2012.)

of 0.05 mA cm^{-2} to 4.35 V. In addition, Shao-Horn and colleagues [48] claimed that DMSO may not be stable for the oxygen electrode in the development of long cycle life rechargeable Li–air batteries. However, they also used a carbon oxygen electrode composed of carbon nanotubes, and the effect of carbon decomposition during the charging process should be considered in addition to the effect of the lithium salt.

2.4.4 ELECTRODE CATALYSTS

The difference between the discharge and charge cell voltage results in the high energy loss for the battery. The Li/LiCF$_3$SO$_3$ in TEGDME/Super P carbon/O$_2$ cell exhibited excellent cyclic performance at 500 mA g^{-1} (0.5 mA cm^{-2}), but the difference between the discharge and charge cell voltage was as high as 1.7 V, which corresponds to an energy loss of ca. 50% [60]. Bruce and colleagues [66] reported interesting results for the lithium–oxygen cell with LiClO$_4$ in DMSO. They showed that incorporation of a redox mediator, such as tetrathiafulvalene (TTF), enables recharging at a high rate. Figure 2.12 shows the cyclic stability of the cell (VII):

$$\text{Li/1 M LiClO}_4 \text{ in DMSO-10 mM TTF/nanoporous gold tube/O}_2 \qquad \text{(VII)}$$

The TTF molecule is oxidized to TTF$^+$ at the cathode and then, in turn, Li$_2$O$_2$ is oxidized and in the process, TTF$^+$ is reduced back to TTF. By using a molecular electron–hole transfer agent, the oxidation of Li$_2$O$_2$ is far more effective than possible in its absence.

Many catalysts have been proposed for nonaqueous lithium–air batteries, but most of the early studies used carbonate-based electrolytes, where decomposition of the carbonate electrolyte occurred, in addition to decomposition of the reaction

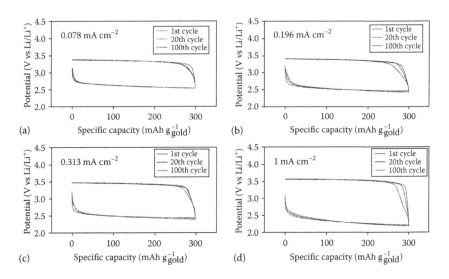

FIGURE 2.12 Cyclic stability of the Li/LiClO$_4$ in DMSO with 10 mM TTF/nanoporous gold electrode/O$_2$ cell. (a) 0.079, (b) 0.196, (c) 0.313, and (d) 1 mA cm^{-2}. (From Chen, Y. et al., *Nat. Chem.*, 51, 489, 2013.)

products rather than the formation and decomposition of Li_2O_2. Luntz and colleagues [67] investigated the catalytic activity of Au, Pt, and MnO_2 nanoparticles for the ORR and OER using the ether based DME electrolyte, and reported a reduction in the overpotentials for both the ORR and OER, where none of these performed better than the Vulcan XC 72 carbon electrode. Black et al. [68] demonstrated that the nanocrystalline Co_3O_4 and reduced graphene oxide in the KB air electrode could significantly reduce the charge overpotential at the initial cycle, although the cyclic performance was not excellent due to the deposition of side reaction products from parasitic reaction with the 1 M $LiPF_6$ in TEGDME electrolyte.

2.4.5 Further Problems for Nonaqueous Lithium–Air Batteries

Most experiments to date on rechargeable lithium–air batteries have used pure oxygen to avoid unwanted parasitic reactions with H_2O and CO_2 from the air. For EVs and stationery energy storage applications, lithium–air batteries should be operated in an ambient atmosphere. However, lithium metal is easily corroded by moisture in the atmosphere, which results in a low reversible capacity and leads to cell failure. Kuboki et al. [49] reported the discharge performance of a lithium–air cell with a hydrophobic ionic liquid under open air at 0.01 mA and room temperature. A steady-state cell voltage was observed for 56 days and then the cell voltage suddenly dropped. Two approaches have been considered to overcome this problem. One approach is to use a water-stable solid lithium-conducting electrolyte, which is introduced in the next section of this chapter. The other approach is to use an oxygen-permeable selective membrane. Zhang et al. [69,70] have proposed several types of oxygen selective membranes to protect the lithium electrode from moisture contamination. A lithium–air cell equipped with a membrane consisting of poly(tetrafluoroethylene) (PTFE) coated on a porous metal substrate sheet as a ca. 5 μm thick dense film at the air cathode interface could operate with a continuous discharge for 21 days in ambient air with 20% relative humidity (RH). However, there has been no perfectly oxygen selective membrane for rechargeable lithium–air batteries to date. This is thus a challenging research subject for practical lithium–air batteries.

The other challenge for rechargeable lithium–air batteries is the suppression of lithium dendrite formation at high current density. Lithium–air batteries should use an anode with lithium because lithium cannot be supplied from the cathode as with lithium–ion batteries. Lithium metal and lithium alloys are better anode candidates for lithium–air batteries because the capacity of the lithium containing compounds proposed for lithium–ion batteries, such as $Li_4Ti_5O_{12}$ [71], is too low to develop a high energy density lithium–air battery. Lithium metal is the best candidate for high energy density lithium–air batteries due to its high specific capacity (3860 mAh g^{-1}) and low negative potential (−3.04 V vs. NHE). However, the lithium metal anode has not been used for rechargeable lithium batteries due to lithium dendrite formation by lithium deposition, which can cause internal short circuiting [72]. In previous reports on nonaqueous lithium–O_2 cells, the formation of dendrites during the charging process was not discussed because the area specific capacity (specific capacity per gram electrode [mAh g^{-1}]×loading of electrode per unit area [g cm^{-2}]) and the charged current density were not so high [45,60]. However, no electrolyte that overcomes dendrite formation at high current density has been reported to date.

2.5 AQUEOUS LITHIUM–AIR SYSTEM

2.5.1 CONCEPT OF AQUEOUS LITHIUM–AIR SYSTEM

Lithium reacts violently with water to produce LiOH and hydrogen gas; therefore, to avoid such parasitic corrosion reactions, most research on lithium–air batteries has employed aprotic electrolytes. However, as shown in the previous section of this chapter, nonaqueous systems have some severe problems that still need to be addressed, such as decomposition of the aprotic solvent (poor cyclic performance), high polarization for the charge and discharge processes (low power density and low energy storage efficiency), and lithium corrosion by water and CO_2 when operated in the air (short shelf-life). However, some of these problems are not applicable to the aqueous system. The aqueous lithium–air system consists of a lithium metal anode, a water-stable lithium ion conducting solid electrolyte, aqueous electrolyte, and a carbon air electrode, as shown in Figure 2.3. This concept was proposed by Visco et al. in 2004 [24], where a NASICON-type lithium ion conducting solid electrolyte of $Li_{1+x}Al_xTi_{2-x}(PO_4)_3$ (LATP) was used. LATP is stable in a high lithium ion concentration aqueous solution [27] but unstable in contact with lithium metal [73]. Therefore, an interlayer between lithium metal and LATP should be used. Visco et al. [24] used a lithium ion conducting solid electrolyte of Li_3N as the interlayer, while Imanishi and colleagues [27] used a lithium conducting solid polymer electrolyte based on PEO. Some reports [74] used a conventional nonaqueous electrolyte, such as 1 M $LiClO_4$ in EC/DEC, but such electrolytes result in significant lithium dendrite formation at high current density and are not applicable for practical-use high energy density lithium–air batteries.

Electrolytes of acidic and basic aqueous solutions have also been studied. The water-stable lithium ion conducting solid electrolyte of LATP is stable in weak acid [75] and weak base [27]. A weak acid solution of $CH_3COOH–CH_3COOLi–H_2O$ [75] and a weak base solution of saturated LiOH with saturated LiCl were proposed by Imanishi and colleagues [26]. The cell reaction of the acetic acid system is as follows:

$$2CH_3COOH + 2Li + 1/2O_2 = 2CH_3COOLi + H_2O \qquad (2.10)$$

The calculated energy density from this reaction is 1500 Wh kg^{-1}, which is lower than that for reaction (2.4) with the weak base electrolyte. The main advantage of operating the lithium–air cell in an acidic solution is that there is no contamination of the electrolyte from CO_2 in the air. The cell reaction for LiOH with LiCl aqueous solution is shown in reaction (2.4). The solubility of the LiOH reaction product is not so high and LiOH is deposited on the carbon electrode and/or in the electrolyte as $LiOH·H_2O$. The calculated energy density of basic lithium–air batteries is 1897 Wh kg^{-1}, which is approximately half of that of nonaqueous lithium–air batteries.

2.5.2 Water-Stable Lithium Ion Conducting Solid Electrolytes

The water-stable lithium ion conducting solid electrolyte is a key material without which the aqueous lithium–air battery could not be operated. The water-stable solid electrolyte for aqueous lithium/lithium–air batteries should have the following characteristics: (1) high lithium ion conductivity, (2) stability in high LiOH concentration aqueous solution, (3) water impermeability, (4) ease of preparation for thin film with excellent mechanical properties, and (5) stability in contact with lithium metal. Various types of solid lithium ion conducting electrolytes have been reported over the last several decades, such as the layered-type structure of Li_3N [76], Li_2S-based glass electrolyte [77], LISICON-type $Li_{14}Zn(GeO_4)_4$ [78], perovskite-type $La_{2/3-x}Li_{3x}TiO_3$ (LLTO) [79], NASICON-type $Li_{1+x}Al_xTi_{2-x}(PO_4)_3$ (LATP) [25], garnet-type $Li_7La_3Zr_2O_{12}$ (LLZ) [80], and $Li_{10}GeP_2S_{12}$ [81]. $Li_{10}GeP_2S_{12}$ reported by Kanno and colleagues [81] has an extremely high lithium conductivity of 1.2×10^{-2} S cm^{-1} at room temperature, which is higher than that of conventional liquid electrolytes because its lithium transport number is unity. However, only LLTO, LATP, and LLZ are presently known to be stable in an aqueous solution with a high concentration of lithium salt.

The water-stable and water-impermeable NASICON-type lithium ion conducting $Li_{1+x+y}Al_x(Ti,Ge)_{2-x}P_{3-y}Si_yO_{12}$ glass ceramics are supplied by Ohara Ltd., Japan, and have an electrical conductivity of $1–3.3 \times 10^{-4}$ S cm^{-1} at room temperature. The stability of the Ohara plate (O-LATP) in aqueous solution has been extensively studied by Imanishi and colleagues [27,75,82,83]. The glass ceramic decomposes in strong acid solution and dissolves in strong alkaline solution. The reaction product for aqueous lithium–air batteries is LiOH and the aqueous electrolyte is saturated with LiOH after approximately 30% discharge depth. Therefore, the lithium conducting solid electrolyte for aqueous lithium–air batteries should be stable in LiOH saturated aqueous solution. Shimonishi et al. found that by immersing O-LATP in alkaline solution, the high resistance phase was segregated at the grain boundary and the electrical conductivity decreased significantly, but O-LATP was stable in a solution saturated with both LiCl and LiOH [27]. The stability of O-LATP in aqueous solution is dependent on the pH and the Li$^+$ concentration. Table 2.2 shows the conductivity of O-LATP immersed in LiOH-LiCl aqueous solution at 50°C for 3 weeks. The O-LATP is stable in the aqueous solution saturated with LiOH and with a high LiCl content. These results suggest that the water-stable O-LATP could be used as a protective layer for the lithium metal anode of lithium–air batteries with a LiCl saturated aqueous solution. O-LATP plates are prepared by quenching from the melt and crystallized at high temperature. The electrical conductivity of the typical water-impermeable O-LATP plate with excellent mechanical properties is around 1×10^{-4} S cm^{-1}. Imanishi and colleagues [84] reported that a composite film of $Li_{1.4}Al_{0.4}Ti_{1.6}(PO_4)_3$–3 wt% TiO_2 and epoxy resin had a high electrical conductivity of 4×10^{-4} S cm^{-1} at 25°C and was water-impermeable. The LATP composite films were prepared by a tape-casting method. The tape-cast LATP-epoxy resin composite film was prepared by dropping a mixed solution of 2,2-bis(4-glycidyloxyphenyl) propane and 1,3-phenylenediamine in THF on the surface of the LATP film. The LATP film with the solution was stored under vacuum for several minutes to allow

TABLE 2.2

Electrical Conductivity of the Ohara LATP Plate in LiCl–LiOH–H₂O at 50°C for 3 Weeks

LiOH Concentration (M)	LiCl Concentration (M)	pH	Conductivity at 25°C ($\times 10^{-4}$ S cm^{-1})
0.57	0	12.50	0.27
0.057	0	12.78	0.58
0.0057	0	11.91	1.7
0.00057	0	10.91	2.0
3.6	8.1	9.36	2.51
5.0	11.8	8.25	2.61
5.1	11.6	8.14	2.32
2.3	10.4	7.62	2.55
0.18	12.6	7.01	3.11

Source: Shimonishi, Y. et al., *J. Power Sources*, 196, 5128, 2011.

penetration of the solution into the pores before being polymerized at 150°C for 24 h. The epoxy resin content was ca. 1.5 wt%. The stability of the LATP-epoxy resin composite film in a solution saturated with LiOH and LiCl was tested using a Li/PEO$_{18}$LiTFSI/LATP-epoxy resin composite film/saturated LiOH and LiCl/Pt, air cell. Figure 2.13 shows the impedance profiles for the cell at 60°C, where a polymer

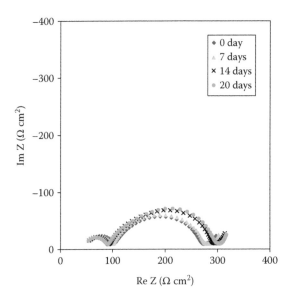

FIGURE 2.13 Impedance profiles at 60°C for the Li/PEO$_{18}$LiTFSI/tape-cast LATP-3 wt%TiO$_2$-epoxy composite film/saturated LiOH and LiCl aqueous solution/Pt, air cell for various storage periods. (From Takahashi, K. et al., *J. Electrochem. Soc.*, 159, A1065, 2012.)

electrolyte of PEO-LiTFSI-10 wt% $BaTiO_3$ was used to protect from direct contact between lithium metal and LATP because LATP is unstable with lithium metal. The impedance profiles show two semicircles; the small semicircle at high frequency corresponds to the contribution of the grain boundary resistance of the LATP composite electrolyte, and the large semicircle corresponds to the contribution of the interface resistance between lithium metal and the polymer electrolyte and the charge transfer resistance. The diameter of the semicircle at low frequency was increased slightly during the initial 7 days and then remained unchanged for another 13 days. Therefore, the LATP-epoxy composite electrolyte is stable and water-impermeable in the saturated LiOH and LiCl aqueous solution.

The specific area resistance of the 150 μm thick O-LATP plate is 150 Ω cm². To improve the resistance of the solid electrolyte in the aqueous lithium–air cell, an electrolyte with conductivity above 10^{-3} S cm^{-1} is desirable. High lithium ion conductivity was reported for the $Li_{1.5}Al_{0.5}Ge_{1.5}(PO_4)_3$ glass ceramic solid electrolyte by Thokchem and Kumar [85] as high as 4.62×10^{-3} S cm^{-1} at 27°C. However, the method for preparation of the glass ceramic was somewhat complex and germanium is an expensive element. Recently, Imanishi and colleagues [82] reported high lithium ion conductivity for the Ge-substituted NASICON-type $Li_{1.4}Al_{0.4}Ge_xTi_{1.6-x}(PO_4)_3$ lithium ion conductors. The highest steady-state conductivity of 1.7×10^{-3} S cm^{-1} at 25°C was observed at x=0.2 in $Li_{1.4}Al_{0.4}Ge_xTi_{1.6-x}(PO_4)_3$. The electrical conductivity was slightly decreased in the ambient atmosphere due to the ion exchange reaction with Li$^+$ in the electrolyte and H$^+$ in air. The stability of the high conductivity electrolyte in various aqueous solutions was examined. Figure 2.14 shows impedance profiles for $Li_{1.4}Al_{0.4}Ge_{0.2}Ti_{1.4}(PO_4)_3$ (LAGTP) immersed in distilled water, saturated LiCl aqueous solution, saturated LiOH and LiCl, and saturated LiOH aqueous solutions

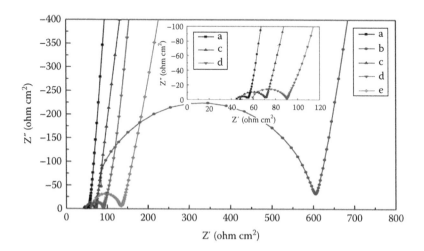

FIGURE 2.14 Impedance profiles at 25°C for (a) pristine $Li_{1.4}Al_{0.4}Ge_{0.2}Ti_{1.4}(PO_4)_3$ and $Li_{1.4}Al_{0.4}Ge_{0.2}Ti_{1.4}(PO_4)_3$ immersed in (b) water, (c) saturated LiCl aqueous solution, (d) saturated LiOH and LiCl aqueous solution, and (e) saturated LiOH aqueous solution at 50°C for 1 week. (From Zhang, P. et al., *Solid State Ionics*, 253, 175, 2013.)

at 50°C for 1 week. The XRD patterns of all samples immersed these solutions for 1 week showed no change with that of the pristine LAGTP. The electrical conductivity of LAGTP immersed in the saturated LiOH and LiCl aqueous solution was comparable to that of LAGTP after 1 week aging in a glove box. It could be concluded that LAGTP with high lithium ion conductivity is stable in the saturated LiOH and LiCl aqueous solution, and is thus acceptable for use as the protective layer for a lithium metal electrode in lithium–air batteries.

The other candidate for the water-stable lithium conducting solid electrolyte is the garnet-type $Li_7La_3Zr_2O_{12}$ (LLZ). In 2007, Weppner and coworkers [80] reported that the garnet-type $Li_7La_3Zr_2O_{12}$ solid electrolyte with lithium excess exhibits a high lithium ion conductivity of 2.44×10^{-4} at 25°C, which is comparable or lower than that of the NASICON-type LATP lithium conductor; however, LLZ is stable in molten lithium metal and is not so sensitive to moisture. This lithium ion conducting solid electrolyte is quite attractive as the protective layer for a water-stable lithium electrode in lithium–air batteries. Doping of LLZ with aluminum is considered to stabilize the high-temperature cubic phase [86]. LLZ is generally prepared at high temperatures above 1100°C using an alumina crucible, where aluminum ions diffuse into LLZ. Imanishi and colleagues [87] reported 0.5 wt% aluminum contamination for LLZ sintered at 1180°C for 32 h. The LLZ reported by Weppner and coworkers [80] was prepared at 1230°C for 36 h using an alumina crucible and may also contain some aluminum. Similar stabilization of the high-temperature cubic phase of $Li_7La_3Zr_2O_{12}$ was also achieved by substitution of Zr^{4+} with Ta^{3+} in the absence of Al [88]. Goodenough and colleagues [89] reported the lithium ion conductivity of $Li_{7-x}La_3Ta_xZr_{2-x}O_{12}$ prepared by the conventional solid state reaction method at 1120°C–1230°C in an alumina crucible, where the highest total conductivity was 1×10^{-3} S cm^{-1} at 25°C for $Li_{6.4}La_3Ta_{0.6}Zr_{1.4}O_{12}$ sintered at 1140°C for 16 h. The stability of Ta-doped LLZ (Ta-LLZ) with lithium metal and in saturated LiOH and LiCl aqueous solution was examined by Imanishi and colleagues [90]. Figure 2.15a shows room temperature impedance profiles for the Pt/saturated LiOH with 10 M LiCl aqueous solution/$Li_{6.75}La_3Ta_{0.25}Zr_{1.75}O_{12}$/saturated LiOH with 10 M LiCl aqueous solution/Pt cell as a function of the storage period. The cell resistance decreased during the first 1 week and then remained unchanged with the storage time. The conductivity calculated from the cell resistance was comparable to that measured using the Au/$Li_{6.75}La_3Ta_{0.25}Zr_{1.75}O_{12}$/Au cell. No significant change was observed in the impedance profiles measured for the Li/$Li_{6.75}La_3Ta_{0.25}Zr_{1.75}O_{12}$/Li cell with a storage period of 3 months, as shown in Figure 2.15b. These results suggest that Ta-LLZ with high lithium ion conductivity is stable in an aqueous solution saturated with both LiOH and LiCl and in contact with lithium metal. Ta-LLZ is thus attractive as the separator for the water-stable lithium metal electrode. The other requirement of the solid electrolyte for the water-stable lithium electrode is reversible lithium stripping and deposition at a high current density. Figure 2.16 shows the polarization behavior for the Li/$Li_{7-x}La_3Ta_xZr_{2-x}O_{12}$/Li cell at 0.5 mA cm^{-2} and 25°C, where the relative density of $Li_{7-x}La_3Ta_xZr_{2-x}O_{12}$ was ca. 96%. An abrupt drop in cell voltage after polarization for a short period was observed for all cells, which may be due to short-circuiting by the formation of lithium dendrites. Similar abrupt cell voltage drops were observed for Li/Al_2O_3-doped LLZ/Li [87] and Li/$Li_{6.75}La_3Nb_{0.25}Zr_{1.75}O_{12}$/Li

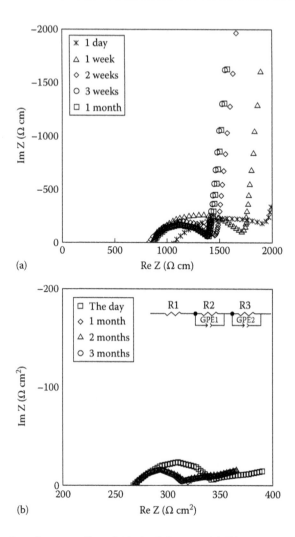

(a)

(b)

FIGURE 2.15 Impedance profiles of (a) the Pt/saturated LiOH and 10 M LiCl aqueous solution/Li$_{6.75}$La$_3$Zr$_{1.75}$Ta$_{0.25}$O$_{12}$/saturated LiOH and 10 M LiCl aqueous solution/Pt cell at room temperature and (b) of the Li/Li$_{6.75}$La$_3$Zr$_{1.75}$Ta$_{0.25}$O$_{12}$/Li cell at 25°C for various storage periods. (From Ishiguro, K. et al., *J. Electrochem. Soc.*, 161, A668, 2014.)

cells [91]. Monroe and Newman [92] predicted that if a homogenous solid electrolyte with a modulus of 6 GPa was obtained, then the lithium dendrite problem would be solved. The modulus of ceramic materials is generally higher than 6 GPa; therefore, we should expect no lithium dendrite formation would occur between LLZ and lithium metal. However, the Li/Li$_{6.75}$La$_3$Ta$_{0.25}$Zr$_{1.75}$O$_{12}$/Li cell showed a short-circuit after 100 s polarization at 0.5 mA cm^{-2}. The mechanism for dendrite formation between lithium metal and LLZ has yet to been clarified, but the lithium diffusion kinetics at grain boundaries may be a key factor because Li$_{6.3}$La$_3$Ta$_{0.7}$Zr$_{1.3}$O$_{12}$ with high grain boundary resistance showed a longer period until short circuit than

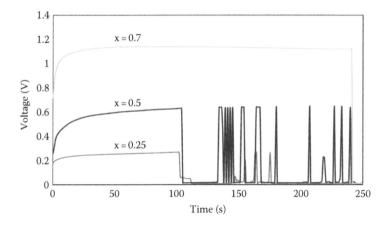

FIGURE 2.16 Cell voltage versus polarization period curves for Li/LLZ-xTa/Li at 0.5 mA cm^{-2} and 25°C. (From Ishiguro, K. et al., *J. Electrochem. Soc.*, 161, A668, 2014.)

$Li_{6.75}La_3Ta_{0.25}Zr_{1.75}O_{12}$ with lower grain boundary resistance and a slightly higher relative density. Therefore, the grain boundary resistance of the water-stable and high-conductivity Ta doped LLZ should be increased to protect from lithium dendrite formation.

2.5.3 PROTECTIVE INTERLAYER BETWEEN WATER-STABLE SOLID ELECTROLYTE AND LITHIUM METAL

At present, the candidates for the protective layer between lithium metal electrode and aqueous electrolyte are only NASICON-type $Li_{1+x}Al_xTi_{2-x}$ $(PO_4)_3$ and perovskite-type $La_{2/3-x}Li_{3x}TiO_3$, but these compounds are unstable in contact with lithium metal. Therefore, an interlayer material is required between lithium metal and the lithium conducting solid electrolyte for the water-stable lithium electrode. The requirements for the interlayer are (1) high lithium ion conductivity, (2) stability in contact with lithium, (3) dendrite formation-free, and (4) ease of thin film preparation. Li_3N [93], Li_{3-x} $P(N,O)_4$ (LiPON) [94], and a lithium conducting polymer electrolyte [95] have been used as interlayer materials. In some reports, the conventional nonaqueous electrolytes have been used, but these electrolytes exhibit serious lithium dendrite formation during the lithium deposition process, which results in a short circuit of the cell [72]. The preparation methods for Li_3N and LiPON are somewhat complicated and fabrication of large-size films is difficult because they are typically deposited on the solid electrolyte using a vapor deposition method. The polymer electrolytes are prepared by a conventional technique; however, the room temperature conductivity is too low and the cell should be operated at higher temperatures above 50°C.

Imanishi and colleagues [95] have extensively studied the PEO-based polymer electrolytes as the interface layer for the water-stable lithium electrode. There are some problems for a PEO-based electrolyte interlayer due to the high interface

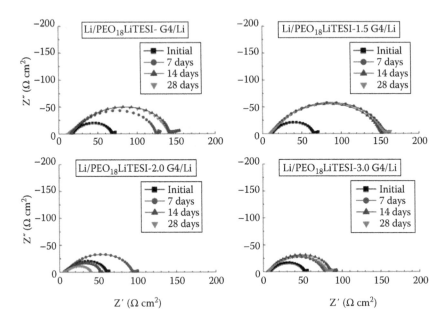

FIGURE 2.17 Impedance spectra for Li/PEO$_{18}$LiTFSI-xTEGDME/Li for various storage times at 60°C. (From Wang, H. et al., *Membranes*, 3, 298, 2013.)

resistance between lithium metal and the interlayer, and lithium dendrite formation. The cell resistance of Li/PEO$_{18}$LiTFSI/Li was reported to be as high as 420 Ω cm^{-2} at 60°C [95]. The cell resistance was significantly decreased to 130 Ω cm^2 at 60°C by the addition of an ionic liquid, N-methyl-N-propylpiperidinium-bis(trifluoromethansulfonyl)imide (PP13TFSI). In addition, the cell resistance was decreased to 85 Ω cm^2 at 60°C by the addition of 2 mol% TEGDME to PEO$_{18}$LiTFSI [96]. Figure 2.17 shows impedance spectra for Li/PEO$_{18}$LiTFSI-xTEGDME/Li as a function of the storage time at 60°C. Low and steady cell resistance was observed for the PEO$_{18}$-2.0TEGDME and PEO$_{18}$-3.0TEGDME electrolytes, of which the lithium plating and stripping overpotentials were as low as 0.3 and 0.4 V at 4.0 mA cm^{-2} and 60°C, respectively.

Lithium dendrite formation during lithium deposition at the interface between lithium and Li$_3$N (or LiPON) has not been reported in detail. The mechanism for lithium dendrite growth in Li/polymer electrolyte/Li has been extensively studied by Brissot and Rosso et al. with the help of a direct *in situ* observation technique and simulations of cell potential evolution [97–99]. They observed the start of dendrite formation at a time that followed a power law as a function of the current density, which was very close to Sand's law [97]. At a low current density of 0.05 mA cm^{-2}, no dendrite formation was observed in Li/PEO$_{20}$LiTFSI/Li for 100 h at 80°C, while at a high current density of 0.7 mA cm^{-2}, dendrites were observed in less than 1 h of the deposition process. For the practical application of a protective polymer electrolyte in a water-stable lithium electrode, lithium deposition should be able to occur at a high current density for a long period without dendrite formation. The observed

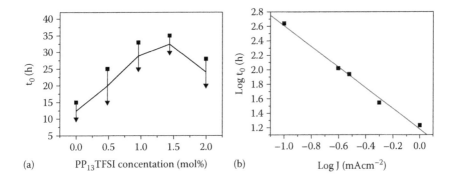

FIGURE 2.18 (a) Dendrite formation onset time (t_0) versus x in $Li_{18}LiTFSI$-xPP13TFSI at 60°C and 0.5 mA cm^{-2}; (▼) no dendrite and (■) dendrite and (b) log current density (J) versus log t_0 for Li/Li$_{18}$LiTFSI-1.44PP13TFSI/Li at 60°C. (From Liu, S. et al., *J. Electrochem. Soc.*, 157, A1092, 2010.)

dendrite formation onset time of 1 h at 0.7 mA is too short. The total weight of lithium for 1 h deposition at 0.7 mA cm^{-2} and with a 10 μm thick copper film substrate is 9.14 mg cm^{-2}, which corresponds to 76.5 mAh g^{-1}. The specific weight capacity is too low to use as an anode for lithium–air batteries. Imanishi and colleagues [100] have reported that the lithium dendrite formation in PEO$_{18}$LiTFSI was suppressed significantly by the addition of PP13TFSI up to x = 1.44 in PEO$_{18}$LiTFSI-xPP13TFSI. Figure 2.18a shows lithium dendrite formation onset time versus x curves at 60°C and 0.5 mA cm^{-2}, where the dendrite formation onset times were observed using a visualization cell. The longest onset time of 35 h was observed for PEO$_{18}$LiTFSI-1.44PP13TFSI. The onset time dependence on the current density is shown in Figure 2.18b. The onset time was decreased with an increase in the current density; that at 0.1 mA cm^{-2} was 434 h and that at 1.0 mA cm^{-2} 17 h. The specific weight capacity including the weight of the 10 μm thick copper film substrate was calculated to be 1270 mAh g^{-1} at 1.0 mA cm^{-2} and 2148 mAh g^{-1} at 0.1 mA cm^{-2}. The specific anode capacity of lithium metal with the PEO$_{18}$LiTFSI-1.44 PP13TFSI interlayer is compared with that for a carbon anode with the conventional nonaqueous electrolyte (339 mAh g^{-1} without copper substrate). It is expected that further studies should enable the development of a lithium dendrite-free inter layer between lithium metal and LATP at current densities as high as 10 mA cm^{-2} for lithium–air batteries in EV applications.

2.5.4 Cell Performance of Rechargeable Aqueous Lithium–Air Batteries

In 2010, Imanishi and coworkers [101] reported successful charge and discharge performance using a cell with an aqueous solution of acetic acid and lithium acetate (VIII),

$$Li/PEO_{18}LiTFSI/O–LATP/CH_3COOH–CH_3COOLi–H_2O/C, air \quad (VIII)$$

The cell consisted of a lithium metal anode, a PEO$_{18}$LiTFSI-10 wt% BaTiO$_3$ interlayer, an O-LATP water-stable lithium ion conducting solid electrolyte, CH$_3$COOH-10 wt% H$_2$O with saturated HCH$_3$COOLi, and a carbon air electrode on carbon paper. The O-LATP is stable in the buffer solution. A schematic diagram of the test cell and its cyclic performance at 60°C and 0.5 mA cm^{-2} under 3 atm are shown in Figure 2.19. The OCV of the cell was 3.69 V at room temperature, which remained constant for 1 month. The calculated OCV from Equation 2.10 is 4.07 V. The difference between

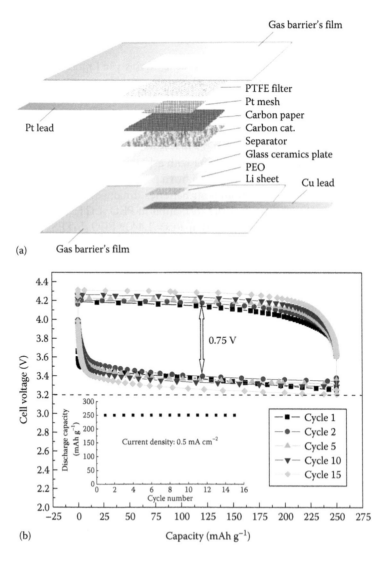

(a)

(b)

FIGURE 2.19 (a) Schematic diagram and (b) charge–discharge cyclic performance of the Li/PEO$_{18}$LiTFSI-10 wt% BaTiO$_3$/O-LATP/CH$_3$COOH–CH$_3$COOLi–H$_2$O/C, air cell at 60°C and 0.5 mA cm^{-2} under 3 atm. (From Zhang, T. et al., *Chem. Commun.*, 46, 1661, 2010.)

the calculated and observed OCVs may be due to a junction potential between the electrolytes and the low activity of the ions. The energy density of the Li/CH$_3$COOH/air system is calculated to be 1320 Wh kg^{-1} from the weight of CH$_3$COOH, lithium, and oxygen and the OCV of 3.69 V, which is approximately 3.3 times higher than that for the graphite/LiCoO$_2$ system. The discharge capacity was calculated as 250 mAh per gram of CH$_3$COOH with consumption of 56% CH$_3$COOH, which corresponds to 680 Wh kg^{-1} using the discharge cell voltage of 3.4 V. Recently, Visco et al. [102] reported a similar aqueous lithium–air battery with malonic acid. The system showed good cyclic performance at a charge current density of 0.5 mA cm^{-2} and a discharge current density of 1.0 mA cm^{-2}. However, high energy density cannot be expected for this system.

Inaguma and Nakashima [103] also studied an aqueous lithium–air cell with the perovskite-type lithium conducting solid electrolyte of La$_{0.57}$Li$_{0.29}$TiO$_3$ (LLTO). The electrical conductivity of the electrolyte is 5.7×10^{-4} S cm^{-1} at 27°C, which is comparable to that of the NASICON-type solid lithium ion conductors. Figure 2.20 shows the cell performance of the cell (IX),

$$\text{Li/1 M LiClO}_4 \text{ in EC–DMC/LLTO/0.5M LiOH aq. solution/}$$
$$\text{porous carbon tube, air,} \qquad \text{(IX)}$$

where a LLTO ceramic tube closed at one end was used as a electrolyte separator (2 mm thick). The stability of LLTO in the aqueous solution was examined. The electrical conductivity of LLTO immersed in 0.5 M LiOH aqueous solution for 4 weeks was slightly decreased to 4.6×10^{-4} S cm^{-1} from 5.0×10^{-4} S cm^{-1}. The OCV was 3.1–3.2 V, which is slightly lower than the theoretical OCV of 3.51 V. Steady discharge cell voltages were observed in the current range from 1 mA (0.06 mA cm^{-2}) to 25 mA (1.6 mA cm^{-2}). The cell voltage drop may be due to the resistance of the LLTO tube. The stability of the LLTO electrolyte in a high concentration of LiOH

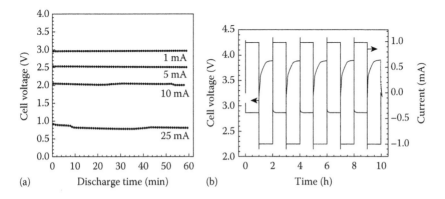

(a)   Discharge time (min)   (b)   Time (h)

FIGURE 2.20 (a) Discharge curves at various currents and (b) cyclic performance for the Li/1 M LiClO$_4$ in EC-DMC/LLTO/0.5 M LiOH/C, air cell at 0.06 mA cm^{-2} and room temperature. The current density is defined with respect to the carbon air electrode surface area of 16 cm^2. (From Inaguma, Y. and Nakanishi, M., *J. Power Sources*, 228, 250, 2013.)

was not examined and is thus questionable. However, LLTO may be stable in saturated LiOH with saturated LiCl aqueous solution, as with the high lithium ion conductivity solid electrolytes of LATP and LLZ.

Stevens et al. [94] demonstrated the rechargeable aqueous lithium–air cell (X)
with saturated LiOH aqueous solution and untreated ambient air:

$$\text{Li/LiPON/O-LATP/saturated LiOH aq. solution/carbon,}$$
$$\text{polymer electrolyte/air} \qquad\qquad\qquad (X)$$

The cell consisted of a lithium electrode, O-LATP with LiPON as the protective
layer between lithium metal and O-LATP, a LiOH saturated aqueous solution, a
composite carbon air electrode with an integrated anionic polymer electrolyte membrane, which prevents lithium carbonate and lithium hydroxide precipitation inside
the air electrode, and a third electrode for the OER to prevent degradation of the
air electrode from oxygen evolution by decoupling it during charge. Figure 2.21
shows a schematic diagram of the aqueous lithium–air cell proposed by Stevens
et al. [94]. The cycling performance at 0.6 mA for 10 min (0.1 mAh cm^{-2}) showed
no degradation for more than 100 cycles; however, the cell cycled at 2 mAh cm^2
was degraded in most cases by a loss of the interface between lithium metal and the
ceramic electrolyte.

The cycling performance of an aqueous lithium–air cell with a high specific
capacity was reported by Imanishi and colleagues [28] in 2014. A schematic diagram of the test cell is shown in Figure 2.22, where the third electrode proposed by

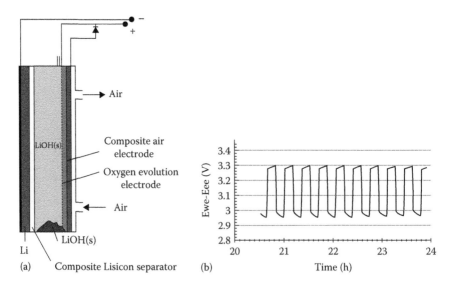

FIGURE 2.21 (a) Schematic diagram of the aqueous lithium–air battery proposed by
Stevens et al. and (b) cyclic performance of Li/LiPON/O-LATP/LiOH aqueous solution/
anionic polymer electrolyte/carbon, air cell with the third oxygen evolution electrode at
0.6 mA cm^{-2} for 10 min. (From Stevens, P. et al., *ECS Trans.*, 28(32), 1, 2010.)

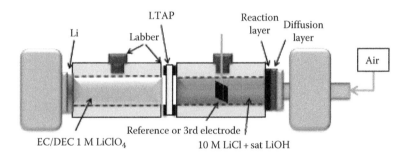

FIGURE 2.22 Schematic diagram of the test cell of Li/1M $LiClO_4$ in EC-DEC/O-LATP/ saturated LiOH with 10M LiCl/KB, air with an OER electrode. (From Sunahiro, S. et al., *J. Power Sources,* 262, 338, 2014.)

Stevens et al. [94] for the OER was applied. RuO_2 on Ti mesh was used as the oxygen evolution electrode. Figure 2.23 shows the KB carbon electrode potential change for the ORR and OER at 0.88 mA cm^{-2} and 25°C in the Li/$LiClO_4$ in EC-DEC/O-LATP/ saturated LiOH with 10 M LiCl aqueous solution/KB, air cell using a platinum with platinum black reference electrode, where the amount of loaded water and KB were 0.305 and 0.0255 g, respectively. The cell was discharged up to 300 mA $g_{H_2O}^{-1}$, which corresponds to a 30% discharge depth for the capacity of the loaded water and to the KB air electrode capacity of 3600 mAh g_{carbon}^{-1}. A steady electrode potential for discharge was obtained. The electrode potential for charging was low at the initial stage, but gradually increased up to 1000 mV. The electrode potentials for discharge at the second cycle were too high and could not discharged at 0.88 mA cm^{-2}. The high overpotential for the OER after polarization for a long period could be explained by the decomposition of carbon in the saturated LiOH aqueous solution with 10M LiCl, as observed by Imanishi and colleagues [104]. CO gas was observed during the OER above overpotentials of 0.2 V on the carbon electrodes of KB, Vulcan XC-72R, and a high surface area acetylene black (AB-S) in the saturated LiOH with 10 M LiCl. Carbon decomposition during the OER was also observed for the nonaqueous system by Bruce and colleagues [105], where carbon decomposition was observed above 3.5 V versus Li/Li^+ in the $LiPF_6$ in TEGDME (or DMSO) electrolyte using an isotopically labeled ^{13}C electrode. Figure 2.23b shows the charge and discharge performance of the Li/1 M $LiClO_4$ in EC-DEC/O-LATP/saturated LiOH with 10 M LiCl/KB, air cell with a platinum with platinum black third electrode for the OER, where the discharge current density was 0.88 mA cm^{-2} based on the KB electrode and the charge current density was 1.0 mA cm^{-2} based on the platinum electrode, and the amount of water and KB loaded were 0.305 and 0.0255 g, respectively. The cell was discharged up to 300 mAh $g_{H_2O}^{-1}$ and then charged. Both steady-state discharge and charge cell potentials were obtained. The specific energy density calculated from the weight of lithium metal, carbon electrode, and water is 770 Wh kg^{-1} using the cell voltage of 3.0 V, which is approximately two times higher than that for conventional lithium–ion batteries. To obtain even higher specific energy density aqueous lithium–air batteries, utilization of the active material

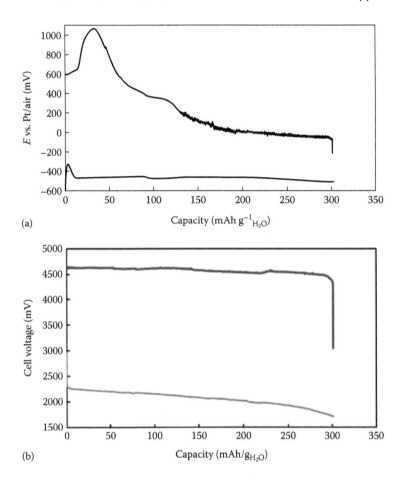

(a)

(b)

FIGURE 2.23 (a) Air electrode potential versus capacity curves for the Li/1 M LiClO$_4$ in EC-DEC/O-LATP/saturated LiOH with 10 M LiCl/KB, air cell at 0.88 mA cm^{-2} and 25°C using a platinum with platinum black reference electrode, and (b) cell potential versus capacity curves for the Li/1 M LiClO$_4$ in EC-DEC/O-LATP/saturated LiOH with 10 M LiCl/KB, air cell using a platinum with platinum black OER electrode at 0.88 mA cm^{-2} for discharge and at 1.0 mA cm^{-2} for charge at 25°C. (From Sunahiro, S. et al., *J. Power Sources*, 262, 338, 2014.)

of water (30% in the result of Figure 2.23b) should be improved to at least 50% and the capacity of carbon electrode (3660 mAh g$_{carbon}^{-1}$) to 6000 mAh g$_{carbon}^{-1}$, where the calculated specific energy density of 1230 Wh kg^{-1} could be estimated and the practical batteries with a specific energy density of more than 300 Wh kg^{-1} could be expected. Imanishi and colleagues [28] have searched for OER electrodes using an aqueous solution of saturated LiOH with 10 M LiCl, such as RuO$_2$, NiCoO$_4$, MnO$_2$, and La$_{0.6}$Sr$_{0.4}$Co$_{0.2}$Fe$_{0.8}$O$_3$, and RuO$_2$ was found to be the best candidate for the OER electrode. RuO$_2$ has high metallic conductivity at room temperature and titanium-coated RuO$_2$/TiO$_2$ is well known as a dimensionally stable electrode for industrial

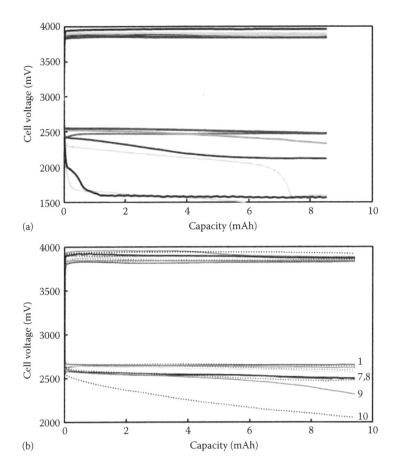

FIGURE 2.24 Cyclic performance for the Li/1 M LiClO$_4$ in EC-DEC/O-LATP/saturated LiOH with 10 M LiCl aqueous solution/KB cell at 0.64 mA cm^{-2} and 25°C (a) in air and (b) in oxygen. (From Sunahiro, S. et al., *J. Power Sources*, 262, 338, 2014.)

chlorine production cells [106]. Byon and colleagues [107] reported excellent catalytic activity of RuO$_2$ dispersed on carbon nanotubes for the OER in a nonaqueous lithium–oxygen cell. The long-term stability of the cell (XI),

Li/1 M LiClO$_4$ in EC-DEC/O-LATP/saturated LiOH
with 10 M LiCl/KB, air (or O$_2$) cell   (XI),

with the RuO$_2$ OER electrode was tested. Figure 2.24a shows the cyclic performance of the cell at 0.64 mA cm^{-2} and 25°C, where 0.0043 g KB was loaded on the air electrode and the cell was discharged to the specific capacity of the air electrode at 2000 mAh g$_{cathode}^{-1}$, and an excess amount of water was loaded. Steady-state cell voltages for the charge and discharge processes were obtained for the first three cycles; however, at fifth cycle, the discharge cell voltage decreased significantly to 1.57 V after discharge for 7 h mAh (14 h discharge and 75 h total operation time). A steady discharge cell voltage of 1.57 V was observed with further cycling and the cell voltage for charging was almost the same with cycling. The steady-state cell reaction at 1.57 V was considered to be the following reaction, where the air supply was suppressed by an insoluble material covering the KB electrode:

$$2Li + 2H_2O = 2LiOH + H_2 \tag{2.11}$$

A candidate for this insoluble material is Li_2CO_3, which is produced by the reaction of LiOH and CO_2 in air. The calculated cell voltage at $a_{H_2} = 1$ and 25°C for reaction (2.11) is 1.68 V, and hydrogen evolution was detected during the discharge process using a gas chromatograph. The cyclic performance of the cell was improved by passing air through a tube with soda lime to remove CO_2, which suggests that the degradation of the cell performance could be explained by contamination with CO_2 in the air. The same type of cell was operated under pure oxygen flow and the cycling performance at 0.64 mA cm^{-2} and 25°C is shown in Figure 2.24b. Good cyclic performance was observed until the eighth cycle, after which the discharge voltage was gradually decreased with further cycling. The degradation of the cell performance could be explained by the poor cyclability of the lithium electrode in the 1M LiClO₃ in EC-DEC electrolyte [73,108] because much fine lithium powder was observed in the electrolyte. Thus, the search for lithium dendrite-free electrolytes that are stable at high current density is an important focus of research to realize practical aqueous lithium–air batteries.

2.6   SUMMARY

In this chapter, we have reviewed the recent results on nonaqueous and aqueous lithium–air batteries. In these 10 years after the report by Bruce and coworkers in 2006, many research groups have extensively studied rechargeable lithium–air batteries, especially the nonaqueous system. However, acceptable electrolyte and cathode materials for the nonaqueous system have yet to be established. Many researchers have recognized the instability of the carbonate electrolytes used in early reports with the superoxides produced at the oxygen electrode. The stability of ether and DMSO with the superoxides is questionable; some reports indicated stability while others found the electrolyte decomposed during the discharge process. The stability of these electrolytes may thus be dependent on the discharge and charge conditions. High surface area carbons for the oxygen electrode are essential to obtain a high specific capacity because the reaction product of Li_2O_2 is insoluble in the electrolyte and is deposited on the carbon surface. The high specific area carbon materials are oxidized during the charging process. No alternative acceptable oxygen electrode materials have been reported to date. In addition, protection of the lithium metal electrode from water and oxygen in the air is important for practical systems using air; however, there is no final solution to this problem at present. Lithium dendrite formation should also be reduced during high current density charging. It should be emphasized that the nonaqueous lithium–air system has many important material problems to be solved.

The aqueous system has no serious material problems, except lithium dendrite formation between lithium metal and the interlayer during high current density charging. The problem of carbon decomposition during the charging process could be solved by the use of a third electrode for the OER because the LiOH reaction product is soluble in the electrolyte. However, high specific energy density such as 700 Wh kg^{-1} cannot be expected with the aqueous system because the electrolyte is an active material for the cell reaction and due to the additional weight of the ceramic separator. Lastly, the mechanical stability of the thin film ceramic separator is an important factor that requires further research in the field of aqueous lithium–air batteries.

REFERENCES

1. G. Girshkumar, B. McCloskey, A. C. Luntz, S. Swanson, and W. Wilche, Lithium-air battery: Promise and challenges, *J. Phys. Chem. Lett.*, 1 (2010), 2193–2203.
2. S. Takahashi et al., Report of the Government Industrial Research Institute, Osaka, Japan No. 354 (1979).
3. J. Sudworth and R. Tilley, *The Sodium/Sulfur Battery*, Chapman & Hall, London, U.K. (1985).
4. J. M. Tarascon and M. Armand, Issues and challenges facing rechargeable lithium batteries, *Nature*, 414 (2001), 359–367.
5. T. Sakai, Alloys for nickel metal hydride battery, *Z. Phys. Chem., Bd.*, 187S (1994), 333–346.
6. B. Scrosati, Lithium batteries: From early stage to the future, in *Lithium Batteries: Advanced Technology and Applications* (Eds. B. Scrosati, K. M. Abraham, W. van Schalkwijk, and J. Hassoun), Wiley & Sons, New York (2013).
7. T. Ohzuku and A. Ueda, Solid-state redox reaction of LiCoO$_2$ (R3m) for 4 volt secondary lithium cell, *J. Electrochem. Soc.*, 141 (1994), 2972–2977.
8. S. Yang and H. Knickle, Design and analysis of aluminum/air battery system for electric vehicles, *J. Power Sources*, 112 (2002), 162–173.
9. D. Aurbach, Z. Lu, A. Schechter, Y. Gofer, H. Glzbar, R. Turgeman, Y. Cohen, M. Moshkovich, and E. Levl, Prototype systems for rechargeable magnesium batteries, *Nature,* 407 (2000), 724–727.
10. F. L. Littaner and K. C. Tsai, Corrosion of lithium in aqueous electrolyte, *J. Electrochem. Soc.,* 124 (1977), 850–855.
11. E. L. Littaner and K. C. Tsai, Anodic behavior of lithium in aqueous electrolyte. I. Transient passivation, *J. Electrochem. Soc.*, 123 (1976), 771–776.
12. K. W. Semkow and F. A. Sammells, A lithium oxygen secondary battery, *J. Electrochem. Soc.*, 134 (1987), 2084–2085.
13. N. N. Batalov and E. L. Arkhigov, Investigation aimed at development of a high-temperature lithium/air storage battery, *Power Sources*, 19, Abst. P30 (1988).
14. K. M. Abraham and Z. Jang, Polymer electrolyte based rechargeable lithium oxygen battery, *J. Electrochem. Soc.*, 143 (1996), 1–5.
15. K. M. Abraham and M. Alomgir, Li$^+$-conductive solid polymer electrolytes with liquid-like conductivity, *J. Electrochem. Soc.*, 137 (1990), 1657–1658.
16. J. Read, Characterization of lithium/oxygen organic electrolyte battery, *J. Electrochem. Soc.*, 149 (2002), A1190–A1195.
17. J. Read, Ether-based electrolyte for the lithium/oxygen organic electrolyte battery, *J. Electrochem. Soc.*, 153 (2006), A96–A100.
18. T. Ogasawara, A. Debart, M. Holtzapfel, P. Novak, and P. G. Bruce, Rechargeable Li$_2$O$_2$ electrode for lithium batteries, *J. Am. Chem. Soc.*, 128 (2006), 1390–1393.
19. V.-C. Lu, H. A. Gasteiger, M. C. Parent, V. Chiloyan, and Y. Shan-Horn, The influence of catalysts on discharge and charge voltages of rechargeable Li-oxygen batteries, *J. Electrochem. Soc.*, 13 (2010), A69–A67.

20. B. Kumar and J. Kumar, Cathode for solid-state lithium-oxygen cells: Roles of nasicon glass ceramics, *J. Electrochem. Soc.*, 157 (2010), A611–A616.

21. B. Kumar, J. Kumar, R. Leese, J. P. Fellner, S. J. Rodrigues, and K. M. Abraham, A solid-state rechargeable long cycle life lithium-air battery, *J. Electrochem. Soc.*, 157 (2010), A50–A54.

22. B. Kumar, D. Thomas, and J. Kumar, Space-charge-mediated superionic transport in lithium ion conducting glass-ceramics, *J. Electrochem. Soc.*, 156 (2009), A506–A513.

23. J. Hassoun, F. Croce, M. Armand, and B. Scrosati, *Angew. Chem. Int., Ed.*, 50 (2011), 2999–3002.

24. S. J. Visco, F. Nimon, B. Katz, L. Johne, and M. Cho, Lithium metal aqueous batteries, *Twelfth International Meeting on Lithium Batteries*, Abstract # 53, Nara, Japan (2004).

25. H. Aono, F. Sugimoto, Y. Sadaoka, N. Imanaka, and G. Adachi, Ionic conductivity of solid electrolytes based on lithium titanium phosphate, *J. Electrochem. Soc.*, 137 (1990), 1023–1027.

26. T. Zhang, N. Imanishi, A. Hirano, Y. Takeda, and O. Yamamoto, Stability of Li/polymer electrolyte-ionic liquid composite/lithium conducting glass ceramics in an aqueous electrolyte, *Electrochem. Solid State Lett.*, 14 (2011), A45–A48.

27. Y. Shimonishi, T. Zhang, N. Imanishi, D. Im, D. I. Lee, A. Hirano, Y. Takeda, O. Yamamoto, and N. Sammes, A study on lithium/air second batteries-stability of the NASICON-type lithium ion conducting solid electrolyte in alkaline aqueous solution, *J. Power Sources*, 196 (2011), 5128–5133.

28. S. Sunahiro, M. Matsui, Y. Takeda, O. Yamamoto, and N. Imanishi, Rechargeable aqueous lithium-air batteries with an auxiliary electrode for the oxygen evolution, *J. Power Sources*, 262 (2014), 338–343.

29. Y.-C. Lu, H. A. Gasteiger, E. Crumlin, R. McGrure Jr, and Y. Shao-Horn, Electrocatalytic active studies of select metal surfaces and implication in Li-Air batteries, *J. Electrochem. Soc.*, 157 (2010), A1016–A1020.

30. B. D. McCluskey, R. Schefler, A. Speidel, G. Girishkumar, and A. C. Luntz, On the mechanism of non-aqueous Li-O$_2$ electrochemistry on C and its kinetic overpotentials: Some implications for Li-Air batteries, *J. Phys. Chem. C*, 116 (2012), 23897–23905.

31. Z. Peng, S. A. Freunberger, L. J. Hardwick, Y. Chen, V. Giordani, F. Bardt, P. Novak, D. Graham, J. M. Trascon, and P. G. Bruce, Oxygen reactions in a non-aqueous Li$^+$ electrolyte, *Angew. Chem. Int. Ed.*, 50 (2011), 6351–6355.

32. F. Mizuno, S. Nakanishi, Y. Kotani, S. Yokoishi, and H. Iba, Rechargeable Li-air batteries with carbonate-based liquid electrolytes, *Electrochemistry*, 78 (2010), 403–405.

33. S. A. Freunberger, Y. Che, Z. Peng, J. M. Griffin, L. J. Hardwick, F. Bardt, P. Novak, and P. G. Bruce, Reaction in the rechargeable lithium-O$_2$ battery with alkyl carbonate electrolyte, *J. Am. Chem. Soc.*, 133 (2011), 8040–8047.

34. J. Ufheil, A. Wursig, O. D. Schneider, and P. Novik, Acetone as oxidative decomposition products in propylene carbonate containing battery electrolyte, *Electrochem. Commun.*, 7 (2005), 1380–1384.

35. B. D. McCloskey, D. S. Bethume, R. M. Shelby, G. Girishkumar, and A. C. Luntz, Solvents critical role in nonaqueous lithium-oxygen battery, *J. Phys. Chem. Lett.*, 2 (2011), 1161–1166.

36. V. S. Bryantsev, V. Giordani, W. Walker, M. Blanco, S. Zecevic, K. Sakaki, J. Uddin, D. Addison, and G. V. Chase, Predicting solvent stability in aprotic electrolyte Li-air batteries: Nucleophilic substitution by the superoxide anion radical (O*⁻), *J. Phys. Chem., A*, 115 (2011), 12399–12409.

37. V. S. Bryantsev, J. Uddin, V. Giordani, W. Walker, M. Blanco, D. Addison, and G. V. Chase, The identification of stable solvent for nonaqueous rechargeable Li-air batteries, *J. Electrochem. Soc.*, 160 (2011), A160–A171.

38. Y. Chen, S. A. Freunberger, Z. Peng, F. Bard, and P. G. Bruce, Li-O₂ battery with a dimethylformamide electrolyte, *J. Am. Chem. Soc.*, 134 (2012), 7952–7957.

39. S. A. Freunberger, Y. Chen, N. E. Drewett, L. J. Hardwick, F. Bardt, and P. G. Bruce, The lithium-oxygen battery with ether-based electrolytes, *Angew. Chem. Int. Ed.*, 50 (2011), 8609–8613.

40. C. O. Laoire, S. Mukerjee, E. J. Plichta, M. A. Hendrikson, and K. M. Abraham, Rechargeable lithium/TEGDME/O₂ battery, *J. Electrochem. Soc.*, 158 (2011), A302–A308.

41. Z. Zhang et al., Increased stability toward oxygen reduction products for lithium-air batteries with oligether-functionalized silane electrolytes, *J. Phys. Chem. C*, 115 (2011), 25535–25542.

42. Y. C. Lu, H. A. Gasteiger, and Y. Shao-Horn, Catalytic activity trend of oxygen reduction reaction for nonaqueous Li-air batteries, *J. Am. Chem. Soc.*, 135 (2011), 19048–19051.

43. D. Chalasani and B. L. Luchi, Reactivity of electrolytes for lithium oxygen batteries with Li₂O₂, *ECS Electrochem. Lett.*, 1 (2012), A38–A42.

44. C. O. Laoin, S. Mukerjee, K. M. Abraham, E. J. Plichta, and M. A. Hendrickson, Influence of nonaqueous solvent on the electrochemistry of oxygen in the rechargeable lithium-air battery, *J. Phys. Chem. C*, 114 (2010), 9178–9186.

45. Z. Peng, S. A. Freunberger, Y. Chen, and P. G. Bruce, A reversible and high rate Li-O₂ battery, *Science*, 337 (2012), 563–566.

46. M. M. O. Yhotiyl, S. A. Freunberger, Z. Peng, and P. G. Bruce, The carbon electrode in nonaqueous Li-O₂ cells, *J. Am. Chem. Soc.*, 135 (2012), 494–500.

47. M. J. Trahan, S. Mukerjee, E. J. Plichta, M. A. Hendrickson, and K. M. Abraham, Study of Li-air cells utilizing dimethyl sulfoxide–based electrolyte, *J. Electrochem. Soc.*, 160 (2013), A259–A267.

48. D. G. Kwabi, T. G. Batcho, C. V. Amanchukwu, N. Ortiz-Vitoriano, P. Hammond, C. V. Thompson, and Y. Shao-Horn, Chemical instability of dimethyl sulfoxide in lithium-air batteries, *J. Phys. Chem. Lett.*, 5 (2014), 2850–2856.

49. T. Kuboki, T. Okuyama, N. Ohsaki, and N. Takami, Lithium-air batteries using hydrophobic room temperature ionic liquid electrolytes, *J. Power Sources*, 146 (2005), 766–769.

50. S. Randstrom, G. B. Appetecchi, C. Lagergren, A. Moreno, and S. Passerini, The influence of air and its components on the cathodic stability of N-butyl-N-methylpyrrolidinum bis(trifluoromethanesulfonyl)imide, *Electrochim. Acta*, 53 (2007), 1837–1842.

51. M. Pian, J. Wandt, S. Meinai, I. Buchberger, N. Tsiouvaras, and H. A. Gasteiger, Stability of pyrrolidinium-based ionic liquid in Li-O₂ cells, *J. Electrochem. Soc.*, 161 (2014), A1992–A2001.

52. G. A. Ella et al., An advanced lithium-air battery exploiting an ionic liquid-based electrolyte, *Nano Lett.*, 14 (2014), 6572–6577.

53. T. Zhang and H. Zhou, From Li-O₂ to Li-air batteries: Carbon nanotube/ionic liquid gels with a tricontinuous passage of electrons and oxygen, *Angew. Chem. Int. Ed.*, 51 (2012), 11062–11067.

54. H. Nakamoto, Y. Suzuki, T. Shiotsuki, F. Mizuno, S. Higashi, K. Takechi, T. Asaoka, H. Nishikoori, and H. Iba, Ether-functionalized ionic liquid electrolyte for lithium-air batteries, *J. Power Sources*, 243 (2013), 19–23.

55. S. Higashi, Y. Kato, K. Takechi, H. Nakamoto, F. Mizuno, H. Nishikoori, H. Iba, and T. Asaoka, Evaluation and analysis of Li-air battery using ether functionalized ionic liquid, *J. Power Sources*, 240 (2013), 14–17.

56. S. A. Freunberger, Y. Chen, F. Barde, K. Takeuchi, F. Mizuno, and P. G. Bruce, Nonaqueous electrolyte, in *The Lithium Air Battery: Fundamentals* (Eds. N. Imanishi, A. C. Luntz and P. Bruce), Springer, New York, pp. 23–58 (2014).

57. M. S. Whittingham, History, evolution, and further in energy storage, *Proc. IEEE*, 108 (2012), 1518–1534.
58. J. Qian, W. A. Henderson, W. Xu, P. Bhattacharya, M. Enelhard, O. Borodin, and J.-G. Zhang, High rate and stable cycling of lithium metal anode, *Nat. Commun.*, 6 (2015), 6362.
59. K. Kinoshita, *Electrochemical Oxygen Technology*, John Wiley & Sons, New York (1992).
60. H.-G. Jung, J. Hassoun, J.-M. Park, Y.-K. Sun, and B. Scrosati, An improved high-performance lithium-air battery, *Nat. Chem.*, 4 (2012), 579–582.
61. P. Du, J. Lu, K. C. Lau, X. Luo, J. Barefio, X. Zhang, Y. Ren, Z. Zhang, L. A. Curtiss, Y.-K. Sun, and K. Amine, K., Compatibility of lithium salts with solvent of the non-aqueous electrolyte in Li-O$_2$ batteries, *Phys. Chem. Chem. Phys.*, 15 (2013), 5572–5581.
62. J. Lu, J.-B. Park, Y.-K. Sun, F. Wu, and K. Amine, Aprotic and aqueous Li-O$_2$ batteries, *Chem. Rev.*, 114 (2014), 5611–5640.
63. G. A. Elia, J.-B. Park, Y. B. Sun, B. Scrosati, and J. Hassoun, Role of the lithium salt in the performance of lithium-oxygen batteries: A comparative study, *ChemElectroChem*, 1 (2014) 47–56.
64. A. C. Luntz, B. D. McCloskey, S. Gowda, H. Horn, and V. Viswanthan, Cathode electrochemistry in nonaqueous lithium air batteries, in *The Lithium Air Battery: Fundamentals* (Eds. N. Imanishi, A. C. Luntz, and P. Bruce), Springer, New York, pp. 59–120 (2014).
65. K. Takechi, S. Higashi, F. Mizuno, H. Nishikoori, H. Iba, and T. Shiga, Stability of solvents against superoxide radical species for the electrolyte of lithium-air battery, *ECS Electrochem. Lett.*, 1 (2012), A27–A29.
66. Y. Chen, S. A. Freunberger, Z. Peng, O. Fontaine, and P. G. Bruce, Charging a Li-O$_2$ battery using a redox mediator, *Nat. Chem.*, 51 (2013), 489–494.
67. B. D. McCloskey, R. Sheffler, A. Speidel, D. S. Bethune, R. M. Shelby, and A. C. Luntz, On the efficacy of electrocatalysis in nonaqueous Li-O$_2$ batteries, *J. Am. Chem. Soc.*, 111 (2011), 18036–18041.
68. R. Black, J.-H. Lee, B. Adams, C. A. Mine, and L. F. Nazar, The rule of catalysts and peroxide oxidation in lithium-oxygen batteries, *Angew. Chem. Int. Ed.*, 52 (2013), 392–396.
69. J. Zhang, W. Xu, W. Li, and W. Liu, Air dehydration membrane for nonaqueous lithium-air batteries, *J. Electrochem. Soc.*, 157 (2010), A940–A946.
70. J. Zhang, W. Xu, and W. Liu, Oxygen selective immobilized liquid membranes for operation of lithium-air batteries in ambient air, *J. Power Sources*, 195 (2010), 7438–7444.
71. T. Ohzuku, A. Ueda, and N. Yamamoto, N., Zero-strain insertion materials of Li$_{1/3}$Ti$_{5/3}$O$_4$, *J. Electrochem. Soc.*, 142 (1989), 1431–1435.
72. E. Peled, The electrochemical behavior of alkali and alkaline earth metal in nonaqueous battery system—The solid electrolyte interface model, *J. Electrochem. Soc.*, 126 (1979), 2047–2051.
73. N. Imanishi, S. Hasegawa, T. Zhang, A. Hirano, Y. Takeda, and O. Yamamoto, *J. Power Sources*, 185 (2008), 1392–1387.
74. Y. Wang and H. Zhou, A lithium-air battery with a potential to continuously reduce O$_2$ from air delivering energy, *J. Power Sources*, 195 (2010), 358–361.
75. Y. Shimonishi, T. Zhang, P. Johnson, N. Imanishi, A. Hirano, Y. Takeda, O. Yamamoto, and N. Sammes, A study on lithium/air secondary battery-stability of NASICON-type glass ceramics in acid solution, *J. Power Sources*, 195 (2010), 6187–6191.
76. U. V. Applen, Ionic conducting in Li$_3$N single crystal, *Appl. Phys. Lett.*, 30 (1977), 621–623.
77. H. Wada, A. Menetrierm, A. Lavasseurs, and P. Hagenmuller, Preparation and ionic conductivity of new B$_2$S$_3$–Li$_2$S–LiI glass, *Mater. Res. Bull.*, 18 (1983), 189–192.

78. H. Y.-P. Hong, Crystal structure and ionic conductivity of $Li_{14}Zn(GeO_4)_4$ and other new Li^+ superionic conductor, *Mater. Res. Bull.*, 13 (1973), 117–124.

79. N. Inaguma, J. Chen, M. Ito, T. Nakamura, T. Uchida, and M. Wakihara, High ionic conductivity in lithium lanthanum titanate, *Solid State Commun.*, 86 (1993), 689–693.

80. R. Murugan, V. Tangadurai, and W. Weppner, Fast lithium ion conduction in garnet-type $Li_7La_3Zr_2O_{12}$, *Angew. Chem. Int. Ed.*, 46 (2007), 7778–7781.

81. N. Kamaya et al., A lithium super ionic conductor, *Nat. Mater.* 10 (2011), 682–685.

82. P. Zhang, M. Matsui, A. Hirano, Y. Takeda, O. Yamamoto, and N. Imanishi, Water-stable lithium ion conducting solid electrolyte of the $Li_{1.4}Al_{0.4}Ti_{1.6-x}Ge_x(PO_4)_3$ system (A=x=0-1.0) NASICON-type structure, *Solid State Ionics*, 253 (2013), 175–180.

83. S. Hasegawa, N. Imanishi, T. Zhang, J. Xie, A. Hirano, Y. Takeda, and O. Yamamoto, Study on lithium/air secondary batteries-stability of NASICON-type lithium ion conducting glass-ceramics with water, *J. Power Sources*, 189 (2008), 371–379.

84. K. Takahashi, P. Johnson, N. Imanishi, N. Sammes, Y. Takeda, and O. Yamamoto, A water-stable high lithium ion conducting $Li_{1.4}Ti_{1.6}Al_{0.4}(PO_4)_3$-epoxy resin hybrid sheet, *J. Electrochem. Soc.*, 159 (2012), A1065–A1069.

85. J. S. Thokchem and J. Kumar, Composite effect in super ionically conducting lithium aluminum germanium phosphate based glass ceramics, *J. Power Sources*, 185 (2008), 480–485.

86. E. Rangasamy, J. Wolfenstine, and J. Sakamoto, The role of Al and Li concentration on the formation of cubic garnet solid electrolyte of nominal composition $Li_7La_3Zr_2O_{12}$, *Solid State Ionics*, 208 (2012), 28–32.

87. R. Sudo, Y. Nakata, K. Ishiguro, M. Matsui, A. Hirano, Y. Takeda, O. Yamamoto, and N. Imanishi, Interface behavior between garnet-type lithium-conducting solid electrolyte and lithium metal, *Solid State Ionics*, 262 (2014), 151–154.

88. R. Inada, K. Kusakabe, T. Tanaka, S. Kudo, and Y. Sakurai, Synthesis and properties of Al-free $Li_{7-x}La_3Zr_{2-x}Ta_xO_{12}$ garnet related oxides, *Solid State Ionics*, 261 (2014), 95–99.

89. Y. Li, J.-H. Han, C.-A. Wang, H. Xie, and J. B. Goodenough, Optimizing Li^+ conductivity in a garnet framework, *J. Mater. Chem.*, 22 (2012), 15357.

90. K. Ishiguro, H. Nemori, N. Sunahiro, Y. Nakata, R. Sudo, M. Matsui, Y. Takeda, O. Yamamoto, and N. Imanishi, Ta-doped $Li_7La_3Zr_2O_{12}$ for water-stable lithium electrode of lithium-air batteries, *J. Electrochem. Soc.*, 161 (2014), A668–A674.

91. K. Ishiguro, Y. Nakata, M. Matsui, I. Uechi, Y. Takeda, O. Yamamoto, and N. Imanishi, Stability of Nd-doped cubic $Li_7La_3Zr_2O_{12}$ with lithium metal, *J. Electrochem. Soc.*, 160 (2013), A1690–A1693.

92. C. Monroe and J. Newman, The impact of elastic deformation on deposit kinetics at lithium polymer electrolyte, *J. Electrochem. Soc.*, 152 (2005), A396–A404.

93. S. Visco, E. Nimon, and B. D. Katz, Ionically conductive membranes for protection of active metal anodes and battery cells, US Patent 772228 (2004).

94. P. Stevens, G. Toussainit, G. Caillon, P. Viaud, F. Vinatief, C. Cantau, O. Ficher, C. Sarrazin, and M. Mallouki, Development of a lithium air rechargeable battery, *ECS Trans.*, 28(32) (2010), 1–12.

95. T. Zhang, N. Imanishi, S. Hasegawa, A. Hirano, J. Xie, Y. Takeda, O. Yamamoto, and N. Sammes, Li/polymer electrolyte/water-stable lithium-conducting glass ceramics composite for lithium-air secondary batteries with an aqueous electrolyte, *J. Electrochem. Soc.*, 155 (2008), A965–A969.

96. H. Wang, M. Matsui, Y. Takeda, O. Yamamoto, D. Im, D. J. Lee, and N. Imanishi, Interface properties between lithium metal and a composite polymer electrolyte of $PEO_{18}LiTFSI$-tetraetheylen glycol dimethyl ether, *Membranes*, 3 (2013), 298–310.

97. C. Brissot, M. Rosso, J. N. Chazaliviel, and S. Lascaud, Concentration measurements in lithium/polymer-electrolyte/lithium cell during cycling, *J. Power Sources*, 94 (2001), 212–318.

98. M. Rosso, C. Brissot, A. Teyssot, M. Dolle, L. Sunnier, J.-M. Tarascon, R. Boucher, and S. Lascaud, Dendrite short-circuit and fuse effect on Li/polymer/Li cells, *Electrochim. Acta*, 51 (2006), 5134–5140.

99. M. Rosso, T. Gobron, C. Brissot, J.-N. Chazalviel, and S. Lascaud, Onset of dendrite growth in lithium/polymer cells, *J. Power Sources*, 97/98 (2001), 804–806.

100. S. Liu, S. N. Imanishi, T. Zhang, A. Hirano, Y. Takeda, O. Yamamoto, and J. Yang Lithium dendrite formation in Li/poly(ethylene oxide)-lithium bis(trifluoromethane sulfonyl)imide and N-methyl-N- propylpiperidinum bis(trifluoromethane sulfonyl) imide/Li cells, *J. Electrochem. Soc.*, 157 (2010), A1092–A1098.

101. T. Zhang, N. Imanishi, Y. Shimonishi, A. Hirano, Y. Takeda, O. Yamamoto, and N. Sammes, A novel high energy rechargeable lithium/air battery, *Chem. Commun.*, 46 (2010), 1661–1663.

102. S. J. Visco, V. Y. Nimon, A. Petrov, K. Pridstko, N. Goncharenke, E. Nimon, L. De Jonghe, Y. M. Volfkovich, and D. A. Bograchev, Aqueous and non aqueous lithium-air batteries enabled by water-stable lithium metal electrodes, *J. Solid State Electrochem.*, 18 (2014), 1443–1456.

103. Y. Inaguma and M. Nakashima, A rechargeable lithium-air battery using a lithium ion-conducting lanthanum lithium titanate ceramics as an electrolyte separator, *J. Power Sources*, 228 (2013), 250–255.

104. H. Ohkuma, I. Uechi, M. Matsui, Y. Takeda, O. Yamamoto, and N. Imanishi, Stability of carbon electrode for aqueous lithium-air secondary batteries, *J. Power Sources*, 245 (2014), 947–952.

105. M. M. Ottakam, T. Stefan, S. A. Freunberger, Z. Peng, and P. G.Bruce, The carbon electrode in nonaqueous Li–O$_2$ cells, *J. Am. Chem. Soc.*, 135 (2013), 494–500.

106. S. Trasatti, Electrocatalysis understanding the success of DSA, *Electrochim. Acta.*, 45 (2000), 2377–2385.

107. E. Yilmaz, C. Yogi, K. Yamamaka, T. Ohta, and H. R. Byon, Promoting formation of noncrystalline Li$_2$O$_2$ in the Li–O$_2$ battery with RuO$_2$ nanoparticles, *Nano Lett.*, 13 (2013), 4679–4684.

108. H. F. Park, C. H. Hong, and W. Y. Yoon, The effect of internal resistance on dendrite growth on lithium metal electrode in the lithium secondary batteries, *J. Power Sources*, 178 (2008), 765–768.

3 Aluminum–Air Batteries
Fundamentals and Applications

Fei Ding, Jun Zong, Sihui Wang, Hai Zhong, Qingqing Zhang, and Qing Zhao

CONTENTS

3.1 INTRODUCTION

3.1.1 History

Metal–air batteries (metal=Li, Zn, Al, Mg, Fe, and Ca) have much higher specific energies than most currently available primary and rechargeable batteries [1,2], so they have been attracting many works in the electrochemistry research. Among these systems, the aluminum–air battery (AAB) is one important member with a practical specific energy density of 400 Wh kg^{-1} [1,2]. AAB is also one of the most developed members due to the low cost and abundance of aluminum metal. The AAB is typically used as a primary metal–air battery with an aluminum anode and an air-breathing cathode in contact with an aqueous electrolyte, typically sodium hydroxide, potassium hydroxide, or sodium chloride [3].

A significant amount of work was carried out on AAB in the 1960s and the early 1970s; the related work on AAB in the past 50 years was already summarized by some researchers as shown in Figure 3.1 [1], most of which are accessible as publications and patents, while others (such as internal company reports) are more restricted. The related works mainly focused on the AAB electrochemistry and their developing applications. Great efforts in the field were received to solve the problems associated with the air electrode, thermal management, and the reversibility of the system. Just in 2014, an electric car which was equipped with an AAB system and developed by two companies (Phinergy, United States, and Alcoa, Israel) made its track debut [4]. When used to supplement a lithium-ion battery, the battery can extend the range of an electric car by about 1600 km (994 miles) [4]. The reports give us enough confidence to believe that the AAB has the promising prospects in the applications. The AAB will become the ideal power in the twenty-first century.

3.1.2 AAB CHEMISTRY

In striking contrast to most other normal batteries (not metal–air batteries), AABs as the metal–air batteries are different in that the active cathode material (oxygen) is not stored in the battery. Instead, oxygen can be absorbed from the environment and then reduced by catalytic surfaces in the air electrode. Figure 3.2 [2] shows the typical structure of AAB.

The discharge reaction at the aluminum electrode (involving the oxidation of aluminum to aluminate ions) can be written as follows [1,2]:

$$Al + 4OH^- \rightarrow Al(OH)_4^- + 3e^- \quad E^0 = -2.35 \text{ V vs. Hg/HgO} \tag{3.1}$$

The cathode electrode is an air or gas diffusion electrode, typically comprising a carbon-based structure that makes oxygen and the electrolyte come in contact with the catalytic surfaces, resulting in the reduction of the oxygen [1,2]:

$$O_2 + 2H_2O + 4e^- \rightarrow 4OH^- \quad E^0 = +0.40 \text{ V vs. Hg/HgO} \tag{3.2}$$

The ideal overall discharge reaction of AAB can be expressed as [1,2]

$$4Al + 3O_2 + 6H_2O + 4OH^- \rightarrow 4Al(OH)_4^- \quad E^0 = 2.75 \text{ V} \tag{3.3}$$

So the theoretical voltage of AAB is 2.75 V, but actually, the operating voltage decreases in the range of 1.0–2.0 V. The reasons for Al electrodes operating at a significantly lower voltage can be summarized as follows [1,2]:

1. An oxide layer covers the surface and increases internal resistance, which will cause a delay in reaching a steady-state voltage. Eventually, the solution becomes supersaturated due to the formation of aluminum hydroxide precipitate:

$$Al(OH)_4^- \rightarrow Al(OH)_3 + OH^- \tag{3.4}$$

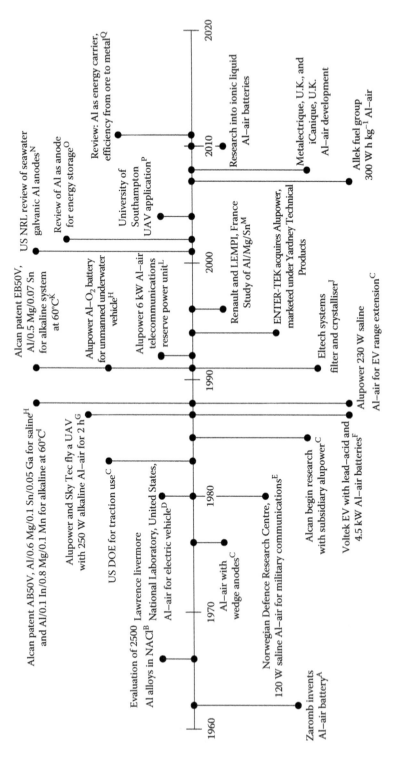

FIGURE 3.1 Timeline of the history of the development of AABs. A–Q are the literature cited in Reference 1. EV, electric vehicle; UAV, unmanned aerial vehicle. (From Egan, D.R. et al., *J. Power Sources*, 236, 293, 2013.)

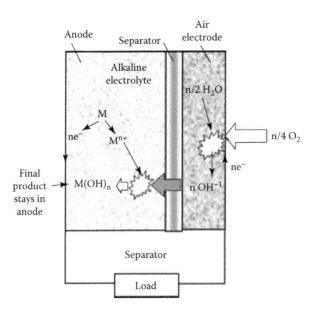

FIGURE 3.2 A schematic figure of AAB. (From Daniel, C. and Besenhard, J.O., *Handbook of Battery Materials*, Wiley-VCH Verlag GmbH & Co. KGaA, Weinheim, Germany, 2011.)

2. Besides that, Al also undergoes a parasitic corrosion reaction, resulting in less utilization of the metal and the formation of hydrogen:

$$2Al + 2OH^- + 6H_2O \rightarrow Al(OH)_4^- + 3H_2 \tag{3.5}$$

3.1.3 AAB BATTERY COMPONENTS

As shown in Figure 3.2, the main components are the air electrode, electrolyte, and aluminum anode.

The air electrode with complex structure is doubtless one of the most important components in the AABs. An effective air electrode is key to the AABs' good performance. Typically, the composite air electrodes are composed of the high-surface-area carbon materials, the catalysts, Teflon binder, air-permeable membranes, and metal mesh current collector. It is divided into three layers: porous catalyst layer, metal mesh current collector, and hydrophobic air-permeable layer. The porous catalyst layer provides the sites for the reduction of oxygen, in which the oxygen diffusing from the environment, the catalyst, and the liquid electrolyte form the three-phase interface (active reaction sites) for electrochemical reaction; and the current collector acts as the support for the electrode and is conductive at the same time; the hydrophobic air-permeable layer can prevent the electrolyte from blocking the gas channels and facilitate oxygen diffusion for the system [1–3]. The details for the air electrode will be concretely introduced in Section 3.3.

Typically, the electrolytes used in AABs are aqueous alkaline solution (such as sodium hydroxide or potassium hydroxide) and neutral saline (such as sodium chloride) as mentioned in Section 3.1.1 [3]. The reason that neutral saline electrolytes are also used is that they exhibit a lower open-circuit corrosion rate and a lower risk to the health of the system compared with caustic solutions. However, the higher conductivity and $Al(OH)_3$ solubility of alkaline solutions can facilitate the delivery of high cell power (175 W kg^{-1}) and energy densities (400 Wh kg^{-1}) compared with saline systems (30 W kg^{-1}, 220 Wh kg^{-1}) [3]. It is attractive for high-power applications such as standby batteries, unmanned underwater vehicles, electric vehicles, and so on. The details for the electrolytes will be concretely introduced in Section 3.4.

Another important component of AABs is the aluminum anode. It is not so simple as that component of the aluminum anode is just the superpure aluminum metal. In fact, the superpure aluminum is unsuitable for use as the anode of AABs, especially in uninhibited alkaline electrolytes. The reasons are as follows: (1) a passive hydroxide layer will form on the surface, resulting in the high overpotential during anodic dissolution; and (2) it also suffers from the high corrosion currents as water is reduced on preferential surface sites evolving hydrogen, which is significantly harmful to the health of the system. One effective method is to alloy the superpure aluminum with other elements (such as Sn, In, Ga) to improve its electrochemical performance [1]. Also, it must be noted that impurities such as copper, iron, and silicon can aggravate self-corrosion. There are also some other effective methods to improve the performance of the anode, which will be concretely introduced in Section 3.2.

The separators of AABs play important roles in separating the two electrodes and allowing normal ion diffusion, ensuring the operation of the system. The details of the separators will be concretely introduced in Section 3.5.

3.1.4 ELECTROCHEMICAL PERFORMANCE

As mentioned in the previous section, AABs have great potential for using as low-cost, high-performance energy-storage systems. The great electrochemical performance is one of the most important things being focused on. As is known, the theoretical specific capacity and energy of a battery are usually calculated on the overall cell reaction (based on the active components of the anode and cathode). In case of AABs, however, we calculated the theoretical specific capacity and energy just based on the electrochemical potential and ampere-hour capacity of the anode because the oxygen is absorbed from the environment (same as fuel cell system). A comparison of different (metal–air batteries) MABs is shown in Table 3.1 [2]. The practical specific energy density and power delivered by AABs with alkaline electrolytes are 400 Wh kg^{-1} and 175 W kg^{-1}, respectively. The AABs with saline systems can deliver 220 Wh kg^{-1} and 30 W kg^{-1}. Therefore, AABs with alkaline electrolytes are more suitable for standby batteries, unmanned underwater vehicles, electric vehicles, and so on. Figure 3.3 [5] shows the discharge characteristics of an AAB using a moderately alkaline electrolyte. The electrochemical performance of AABs will be also concretely introduced in the following sections.

TABLE 3.1

Electrochemical Parameters of the Different Metal–Air Batteries in Aqueous Electrolytes

Battery	Specific Capacity Based on Anode (Ah g⁻¹)	Theoretical Voltage (V)	Practical Voltage (V)	Specific Energy Based on Anode (Wh kg⁻¹)	Specific Energy Based on Anode and H_2O Reactants (Wh kg⁻¹)	Specific Energy Based on Overall Reaction (Wh kg⁻¹)
AAB	2.980	2.71	1.30	3,874	1936	1340
Li/air	3.862	3.45	3.00	11,586	5044	3359
Zn/air	0.820	1.65	1.10	902	902	725
Mg/air	2.205	2.93	1.30	2,867	1647	1195
Fe/air	0.960	1.30	1.00	960	726	597
Ca/air	1.337	3.12	2.00	2,675	1846	1447

Source: Daniel, C. and Besenhard, J.O., *Handbook of Battery Materials*, Wiley-VCH Verlag GmbH & Co. KGaA, Weinheim, Germany, 2011.

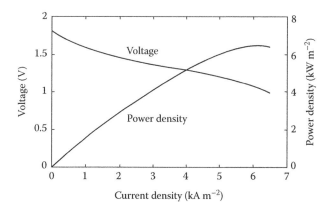

FIGURE 3.3 Discharge characteristics of an AAB using a moderately alkaline electrolyte. (From Yang, S. and Knickle, H., *J. Power Sources*, 112(1), 162, 2002.)

3.1.5 APPLICATIONS

The AAB with good performance can be applied in many fields. But the AABs with different electrolytes will be applied in different fields. For saline systems, compared with the alkaline electrolytes, the anode has a lower corrosion rate, which is healthier for the battery. They have great potential in lower-power and high-energy applications, such as portable devices, settled power sources, ocean buoy, and briny battery [6]. In the briny battery, it can use the oxygen dissolving in the seawater, which can deliver high energy. However, for the alkaline systems, the higher conductivity and

$Al(OH)_3$ solubility of alkaline solutions can facilitate the delivery of higher power. It is greatly attractive for high-power applications such as standby batteries, battlefield power devices (Special Operation Forces Aluminum Air), unmanned underwater vehicles, electric vehicles, and so on [6]. Unmanned underwater vehicles include unmanned submarine, mine sweeping devices, long-range torpedo, and so on [6]. Also note that the major technical gaps of AABs toward the commercialization are the hydrogen evolution and parasitic corrosion reaction of the anode.

3.1.6 Economics and Market

The reports about Phinergy, United States, and Alcoa, Israel, are good evidence to prove that the AAB has promising prospects in the economics and market. For now, though, it seems like that AAB only serves as a range extender and a standard lithium battery is still the primary energy source. Phinergy has also reported that it has signed a contract with a global automaker to bring AAB to the production of cars in 2017, though it did not give a clear statement if AAB would be used as a range extender or as the primary power source. It is a positive signal that the electric vehicles (EVs) equipped with AABs will ship into the market and hold a certain market share in the immediate future. Besides that, the AAB will also hold a certain market share in the smart grid, consumer electronics (such as mobile phone, portable computer, electronic camera), aerospace and defense, and so on.

3.2 AAB ANODES

3.2.1 Introduction

Aluminum is of interest as a battery anode for a number of reasons: (1) its ability to transfer three electrons per atom, (2) its low atomic mass, and (3) its high negative standard potential. These characteristics lead to a high theoretical energy density of 8.1 Wh g^{-1} in AAB. In addition, the large natural abundance and the low production cost of aluminum make it extremely appealing for use in battery systems [7].

Thermodynamically, an aluminum anode should exhibit a potential of 1.66 V in saline and 2.35 V in alkali electrolyte. However, practical aluminum electrodes operate at a significantly lower potential because (a) aluminum is normally covered by an oxide film that causes a delay in reaching a steady-state voltage due to internal resistance and (b) aluminum undergoes a parasitic corrosion reaction, resulting in less than 100% utilization of the metal and the evolution of hydrogen. The discharging reaction of an aluminum electrode in alkaline media is

$$Al + 4OH^- - 3e^- = Al(OH)_4^- \quad E = 2.4\,V \text{ vs. SHE [8]} \tag{3.6}$$

The progressive consumption of hydroxyl ions at the aluminum electrode makes the electrolyte saturated with the aluminate. Eventually, the aluminate concentration exceeds the supersaturation and a crystalline form of aluminum hydroxide precipitates with the regeneration of hydroxyl ions:

$$Al(OH)_4^- = Al(OH)_3 + OH^- \tag{3.7}$$

In addition to the electrochemical consumption of the anode, aluminum is thermo-dynamically unstable in an alkaline electrolyte and reacts with the electrolyte to generate hydrogen:

$$2Al + 6H_2O = 2Al(OH)_3 + 3H_2 \qquad (3.8)$$

This parasitic corrosion reaction, or self-discharge, degrades the coulombic efficiency of the anode and must be suppressed in order to minimize the capacity loss [9].

3.2.2 ANODE TYPES AND DESIGN

Research on anode for AABs has focused on aluminum alloy anode and pure alumi-num anode, which was chosen by the corresponding electrolyte systems.

Most of the anodes used for AABs did not need great efforts for physics design, except the size, which should be appreciated for the cells. But great efforts have been made on anode ingredient design for better cells performance, for example, using alloyed materials as anode. An effective alloying element should possess the following properties: (1) a melting point below the melting temperature of aluminum (657°C), (2) good solid solubility in the aluminum matrix, (3) a higher nobility than aluminum in the electrochemical series as determined from the Pourbaix diagram [10,11], (4) good solubility in an alkaline electrolyte, and (5) a high hydrogen overpo-tential [12]. According to these rules, research on anode alloys for AABs has focused on Mg, Zn, Pb, Sn, Ga, In, Mn, Hg, and Tl alloying elements.

3.2.3 ELECTROCHEMICAL PERFORMANCE

Several studies have investigated the effect of alloying elements on more commer-cially pure aluminum.

3.2.3.1 Binary Aluminum Alloys

3.2.3.1.1 Tin

The dissolution behavior of Al/Sn binary alloys is influenced by the structure, con-centration, and electrochemistry of tin along with the electrolyte temperature. The upper limit for the tin concentration in a binary aluminum alloy for use in AABs is 0.12 wt%, as this is the maximum that can be accommodated in solid solution in the aluminum matrix. A suitable solution heat treatment involves heating at 600°C for several hours, then water quenching [13,14]. An Al/0.12Sn alloy showed the most anodic behavior among a range of Al/Sn binary alloys in a 4 mol dm^{-3} NaOH solu-tion at 25°C (see Figure 3.4). This alloy showed enhanced anodic currents compared to pure aluminum over a potential range corresponding to the region of stability for tin, as discussed earlier [15,16]. Concentrations of tin lower than 0.12% in a 25°C electrolyte most likely had fewer tin deposits formed during the dissolution deposition process to accommodate film-free dissolution of aluminum. The alloys with tin levels greater than 0.12% did not improve the anodic current peak at 1.0 V versus Hg/HgO (see Figure 3.4) because the excess tin was present as second-phase particles on their grain boundaries, which are ineffective at activating aluminum.

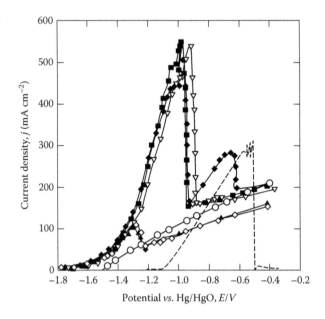

FIGURE 3.4 Effect of tin concentration on the anodic polarization of binary aluminum alloys in alkaline solutions at 25°C. Data from rotating disk, potential sweep experiments performed on six Al/Sn alloys in 4 mol dm^{-3} NaOH [11,16]. Alloys were heat-treated at 600° for 2 hours followed by a water quench. Electrode area: 0.5 cm^2. Potential sweep rate: 5 mV s^{-1}. 20 Hz rotation speed. Counter electrode, platinized titanium strip. ○, 99.995% Al; ▲, Al/0.044 wt% Sn; ◇, Al/0.089 wt% Sn; ■, Al/0.12 wt% Sn; ▽, Al/0.23 wt% Sn; ◆, Al/0.45 wt% Sn; -----, pure Sn. (From Egan, D.R. et al., *J. Power Sources*, 236, 293, 2013.)

The anodic peak at 0.66 V versus Hg/HgO in Figure 3.4 for Al/0.45Sn alloy was due to the oxidation of these second-phase particles, which resulted in pronounced grain boundary attack and potential disintegration of surface grains upon anodic discharge. Since the maximum solubility limit of 0.12% Sn changes when alloying with a ternary element, lower concentrations of tin would be required when forming a ternary or quaternary alloy to achieve a solid solution. The effect of lower tin concentrations is strongly influenced by the electrolyte temperature. At 25°C, they did little to improve the anodic currents observed for pure aluminum (Figure 3.4), whereas at 60°C binary alloys with concentrations ranging from 0.022% up to 0.12%, Sn showed identical performance at 60°C, with Al/0.12Sn exhibiting the highest inhibition and discharge efficiencies, as shown in Table 3.2.

3.2.3.1.2   Gallium
The electrochemistry of Al/Ga alloys in alkaline electrolytes is dependent on the gallium content in the alloy, temperature, and the electrochemistry of gallium with the electrolyte. This is indicated by the marginal differences among the galvanostatic discharge potentials shown in Table 3.2. It was the corrosion behavior that separated the tin concentration temperature and the electrochemistry of gallium.

TABLE 3.2

Open-Circuit Corrosion Characteristics and Discharge Efficiencies of Various Binary Aluminum Alloys in Half-Cell Tests

| Alloy | % Inhibition Efficiency at Open Circuit | % Discharge Efficiency under Galvanostatic Discharge | | | | Electrode Potential during Galvanostatic Discharge | | | |
| | | % φ_{dis} at j in mA cm^{-2} | | | | E vs. Hg/HgO/V | | | |
	% φ_{inh}	50	200	300	600	E_{oc}	E_{50}	E_{200}	E_{600}
Binary alloys									
Tin									
Al/0.022Sn[a]	−597	9	12			−1.78	−1.79	−1.70	
Al/0.044Sn[a]	26	15	45				−1.73	−1.67	
Al/0.089Sn[a]	44	25	62			−1.77	−1.74	−1.62	
Al/0.12Sn[a]	67	35	68			−1.70	−1.73	−1.63	
Gallium									
Al/0.013Ga[a]	−370			58		−1.78			
Al/0.026Ga[a]	−1370			61		−1.85			
Al/0.055Ga[a]	−1627			50		−1.86			
Al/0.10Ga[a]	−4324			7		−1.85			
Al/0.24Ga[a]	−2839			7		−1.89			
Al/2.3Ga[a]	−7036			3		−1.89			
Indium									
Al/0.02In[a]	65	35	63			−1.88	−1.78	−1.68	
Al/0.037In[a]	85	51	65			−1.85	−1.76	−1.72	
Al/0.074In[a]	95	7	40			−1.80	−1.78	−1.71	
Al/0.16In[a]	88	5	23			−1.77	−1.78	−1.72	
Al/0.21In[a]	26	5	21			−1.79	−1.78	−1.70	
Al/0.42In[a]	96	5	21			−1.74	−1.77	−1.70	
Al/0.15Mn[b]	76		55		Passive	−1.69		−1.52	
Al/0.04Fe[b]	−89		40		74	−1.45		−1.39	−1.34
Al/0.81Mg[b]	−1876		4		11				

Source: Egan, D.R. et al., *J. Power Sources*, 236, 293, 2013.

Notes: References [a][16,17], [b][21]; Electrolyte, 4 mol dm^{-3} NaOH at 60°C; Solution heat treatments: 600°C for [a] 2 h and [b] 8 h followed by water quench; φ_{inh} is the inhibition efficiency at open circuit, defined as the percentage difference between the corrosion of the alloy and that of pure aluminum; φ_{dis} is the discharge efficiency.

At an electrolyte temperature of 25°C, gallium contents of 0.055 wt% or higher were required to enhance the anodic currents of pure aluminum, which were reported by J. Hunter [16]. Gallium content of 0.1% Ga shows the best performance of hydrogen evolution inhibition at 50°C (see Figure 3.5) [18]. The corrosion behavior of Al/Ga alloys in an alkaline electrolyte at 60°C is shown in Table 3.2. At open circuit, the corrosion rate of all the binary Al/Ga alloys was extremely high with very negative

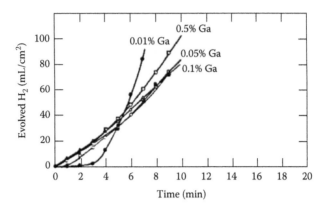

FIGURE 3.5 Hydrogen evolved as a function of time for aluminum–gallium alloys in 4 M KOH solution at 50°C. (From Macdonald, D.D. et al., *Corros. Sci.*, 44(9), 652, 1988.)

inhibition efficiencies. The effect of gallium content on discharge efficiency was dependent on whether the alloy was activated or not, with the activated alloys showing very poor discharge efficiencies, that is, those with gallium levels above 0.1%. It is likely that at temperatures below the low melting temperature of gallium, 29°C, the diffusion of aluminum through solid gallium deposits would be slower and hence the rate of water reduction and corrosion would be less [19].

3.2.3.1.3 Indium

The electrochemistry of Al/In alloys is dependent on the amount of indium in the alloy, the electrolyte temperature, and the electrochemistry of indium [16]. The upper limit for the indium concentration in a binary aluminum alloy for use in AABs is 0.16%, which is close to the solid solubility limit for indium in aluminum at a heat treatment temperature of 640°C. Al/0.16In showed the highest anodic currents in a solution of 4 mol dm^{-3} NaOH at 25°C [16]. Higher indium concentrations showed no further improvement in the anodic behavior, indicating that the polarization behavior was entirely controlled by the indium present in solid solution rather than by second-phase particles. In terms of corrosion behavior of Al/In alloys at 60°C, the data in Table 3.2 show that inhibition and discharge efficiencies for Al/In alloys are dependent on the percentage of indium in solid solution. The Al/In alloys with the lowest indium concentrations showed best effect corrosion/hydrogen evolution inhibition behavior (see Figure 3.6) [18]. In general, the corrosion/hydrogen evolution behavior of Al/In alloys is very complex depending on the discharge time, current density, indium level, and the degree of surface roughening [16].

The anodic polarization curve for Al/0.16%In alloy at 25°C exhibited current fluctuations between 1.3 and 1.05 V versus Hg/HgO, indicating that the alloy was alternating between a more active state and the state exhibited by pure aluminum. These fluctuations could have been caused by successive destruction and build-up of a passive hydroxide layer due to local variations in pH at the active sites [17]. As the aluminum and indium dissolved from the alloy into the electrolyte, the local

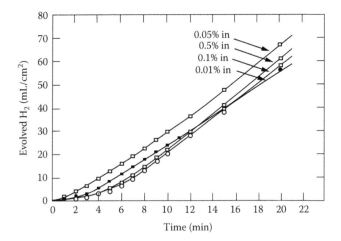

FIGURE 3.6 Hydrogen evolved as a function of time for aluminum–indium alloys in 4 M KOH solution at 50°C. (From Macdonald, D.D. et al., *Corros. Sci.*, 44(9), 652, 1988.)

indate (InO_2^-) concentration increased and eventually reached saturation in the electrolyte. At this point, any further oxidized indium remained on the aluminum surfaces In_2O_3, passivating the active sites. After a brief period of time, when the local indate concentration fell below the solubility limit (by diffusion or convection), the surface In_2O_3 could then dissolve into the electrolyte, and the rate of oxidation of the alloy increased again. This theory explains why the current fluctuations for an Al/Sn alloy (Figure 3.4) were lower than that observed for the Al/In alloys, as SnO_3^{2-} is more soluble in pH 14 media than InO_2^-.

3.2.3.1.4 Other Elements

Examining other binary alloying elements, 0.15% Mn reduced the self-corrosion of 99.99% Al at open circuit, in Table 3.2, and enhanced the anodic behavior with a very negative potential of 1.52 V versus Hg/HgO at 200 mA cm^{-2}. The drawback is that Mn aggravated the parasitic corrosion during discharge, with a discharge efficiency of 55% [21]. Using manganese as an alloying element has been shown to reduce the rate of corrosion of 99.9% aluminum, which contains high levels of iron [22].

The corrosion rate (average) as a function of cold work for pure Al and Al-Zn, Al-Bi, and Al-Te binary alloys in 4 M KOH solution at 50°C is shown in Figure 3.7 [17]. These binary alloys with Bi and Te elements seem to just have a slight improvement on the corrosion/hydrogen evolution inhibition behavior of pure aluminum.

3.2.3.2 Ternary and Quaternary Aluminum Alloys

The pure aluminum and aluminum with additions of magnesium, tin, and/or gallium in 2 M NaCl with a focus on defining alloys that are stable to corrosion on open circuit (i.e., during battery storage) and dissolve anodically at high rate with a high current efficiency was studied by Maria Nestoridi [23]. The stabilities of the materials are compared in Table 3.3, and this table also reports the open-circuit

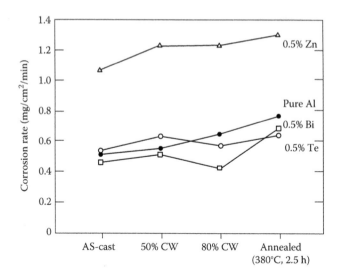

FIGURE 3.7 Corrosion rate (average) as a function of cold work for pure Al and Al-Zn, Al-Bi, and Al-Te binary alloys in 4 M KOH solution at 50°C. (From Macdonald, D.D. et al., *Corros. Sci.*, 44(9), 652, 1988.)

potentials (OCPs) measured in separate experiments (along with their elemental compositions). The current/potential characteristics for a series of the alloys in 2 M NaCl were determined using the controlled current technique following activation with an anodic current density of 1–200 mA cm^{-2}, and the data are reported in Table 3.3. Pure aluminum and the alloys B1 and B3 did not show a negative potential shift during the passage of 50 mA cm^{-2}, and at all current densities, the potential was positive to −800 mV. The alloy B2 (high Sn) showed poorer performance at high current densities, while the alloy B4 (no Mg) gave similar current/potential data in Table 3.3 to other alloys containing Sn and Ga. Hence, magnesium does appear to have a beneficial role in the battery material. Table 3.3 also reports the current/potential characteristics for the three alloys with the same elemental composition (B0, AB50V, and I0) after a heat treatment. Although the differences are small, it appears that the optimum temperature for the heat treatment is 573 K.

3.2.4 ELECTROCHEMICAL TEST PROCEDURES

The anode polarizing curve, cyclic voltammetry curve, and ac resistance diagram were tested using the Electrochemistry Work Station. The three-electrode electrolytic cell system was used under room temperature. The counter electrode was a Pt wire/sheet, the reference electrode was saturated calomel electrode (SCE, Hg/HgO), the work electrode was Al or Al alloy anode, and the electrolyte was aqueous alkaline solution or neutral saline electrolytes.

TABLE 3.3

Composition of Alloys Investigated Together with Open-Circuit Potentials and Current Density/Potential Characteristics for the Alloys in 2 M NaCl at 293 K

Alloy	Open Circuit Potential (mV) vs. SCE	Current Density/Potential vs. SCE (mV)					
		1 mA cm^{-2}	4 mA cm^{-2}	10 mA cm^{-2}	50 mA cm^{-2}	100 mA cm^{-2}	50 mA cm^{-2}
99.999% Al	−800	−775	−770	−750	−660	−530	−230
AlMg3	−820						
AlMg5	−890						
Al/0.5%Mg/0.1%Sn/0.05%Ga	−1530	−1540	−1520	−1480	−1450	−1440	−1390
Al/0.4%Mg/0.1%Sn	−850	−860	−800	−790	−770	−740	−680
Al/0.4%Mg/0.4%Sn/0.03%Ga	−1510	−1530	−1490	−1440	−1380	−1260	−1120
Al/0.4%Mg/0.03%Ga	−800	−810	−790	−760	−720	−700	−620
Al/0.1%Sn/0.03%Ga	−1500	−1530	−1490	−1440	−1380	−1260	−1120
Al/0.6%Mg/0.1%Sn/0.05%Ga	−1530	−1540	−1530	−1510	−1470	−1440	−1380
AB50V							
Al/0.4%Mg/0.07%Sn/0.05%Ga	−1530	−1540	−1530	−1500	−1490	−1480	−1450
IO							
Heated to 873 K for 2 h							
Al/0.5%Mg/0.1%Sn/0.05%Ga		−1530	−1520	−1490	−1480	−1460	−1390
Al/0.6%Mg/0.1%Sn/0.05%Ga AB50V		−1530	−1520	−1500	−1490	−1470	−1400
Al/0.4%Mg/0.07%Sn/0.05%Ga IO		−1550	−1530	−1500	−1490	−1480	−1450

Source: Nestoridi, M. et al., *J. Power Sources*, 178(1), 445, 2008, Tables 1 and 2.

3.2.5 FAILURE MODE ANALYSIS AND MITIGATION STRATEGIES

In an AAB using either alkaline or brine electrolytes, the positive electrode reaction is $O_2 + 2H_2O + 4e^- \rightarrow 4OH^-$ and the negative electrode reaction is $Al - 3e^- \rightarrow Al(III)$, and it is essential to the battery performance sought that both the Al anode and air cathode can operate at a current density 100 mA cm^{-2}. With a neutral brine electrolyte, the Al(III) is largely formed as a solid oxide and/or hydroxide, and the performance of the battery depends critically on the form of this precipitate; it should neither form a passivating film on the aluminum surface nor inhibit the air cathode. In addition, the Al material used as the negative electrode must be stable to corrosion during battery storage, that is, the chemical reactions of $4Al + 3O_2 + 6H_2O \rightarrow 4Al(III) + 12OH^-$ and $2Al + 6H_2O \rightarrow 2Al(III) + 3H_2 + 6OH^-$ should not occur either at the OCP or during anodic discharge (if all the Al is to be converted into electrical energy). It is the competing demands of stability to corrosion and rapid anodic dissolution that make the identification of appropriate aluminum alloys difficult.

To overcome this problem, many researchers have used aluminum alloys of high purity grades possibly doped with elements like Ga, In, Sn, Mg, Mn, and Tl, which act as corrosion inhibitors without increasing the overpotential for aluminum dissolution.

3.2.6 ANODE FABRICATION

Raw materials are commercial pure aluminum, Mg, Zn, Pb, Sn, Ga, In, Mn, Hg, and Tl materials. The nominal composition of the experiment alloys used Al as host materials, and other elements were chosen by requirement. Raw material ingots were cut, dried, and weighed to the required amount of materials and melted in a high-temperature equipment under argon atmosphere at 500°C–800°C. The molten alloy was poured in a preheated cast iron dye. After cold to room temperature, demold and cut it into an appropriate size for batteries requirement.

3.3 AAB CATHODE

3.3.1 INTRODUCTION

The air cathode (often a gas diffusion electrode) is one of the most expensive components of a metal–air battery and is largely responsible for determining the cell performance [24].

The applications of oxygen reduction are strongly dependent on the products involved, either OH$^-$ or HO$_2^-$ [25]. Oxygen reduction is considered to occur mainly through two pathways [26]: (1) a 4e$^-$ reduction reaction without the intermediate formation of hydrogen peroxide and (2) an initial reduction reaction producing H_2O_2, which is possibly then further reduced to OH$^-$. Figure 3.8 shows the schematic illustration of ORR possible pathways.

The actual pathway of oxygen reduction depends on the electrode materials and the electrolyte medium. It is important to mention that in a neutral electrolyte, such as NaCl, O_2 reduction leads to the formation of OH$^-$ and an increase in pH.

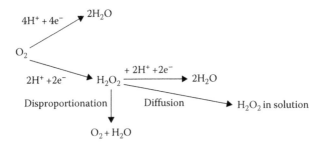

FIGURE 3.8   Schematic illustration of ORR possible pathways.

This is particularly important within the pores of a gas diffusion electrode. This causes an adverse negative shift in the O_2 reduction potential.

3.3.2 Cathode Types (Including Bifunctional Electrodes) and Design

The ORR was one of the limiting factors of metal–air fuel cell in discharging. The cathode reaction is a major factor affecting the performance of metal–air fuel cells, and the electrocatalyst loaded in the cathode is crucial to cathode performance. In principle, most of the catalytic materials applicable to fuel cells could also serve in metal–air batteries and so could be the strategies and techniques to enhance the cathode efficiency. Electrocatalysts for ORR have been extensively studied in fuel cells and Li–air batteries. Noble metal catalysts such as platinum (Pt)-group metals [27–29] have excellent activity and high selectivity, but these catalysts have limited availability and high cost. Therefore, the use of nonprecious metal catalysts is gradually being considered.

The cathode structure used in most metal–air batteries consists of catalyst and Teflon supported on a hydrophobic film with the current collector. The hydrophobic film prevents the seepage of electrolyte from the cell and contributes a fast and uniform supply of air (oxygen) into the cell. Modern air electrode consists of activated carbon porous structure with catalyst such as cobalt and hydrophobic film polytetrafluoroethylene (PTFE). Numerous studies have been conducted on nonprecious metal catalysts, as can be seen in Table 3.4 [30–44].

3.3.3 Electrochemical Performance

The AAB versatility for a wide range of applications depends on what electrocatalyst is selected.

3.3.3.1 Noble Metals and Alloys

Noble metals such as the Pt-group metals, in particular Pt itself, have been intensively studied as ORR catalysts for many decades and continue to spur ongoing interest because of their high stability and superior electrocatalytic activity. Cheaper precious metals such as palladium and silver and their alloys have also been the

TABLE 3.4
Survey of High-Activity Gas Diffusion Electrode Designs

Gas Diffusion Layer Perovskite Type	Catalyst	Active Layer Composition	References
CB, 4.7 m² g⁻¹, 30% PTFE	1.9 mg cm⁻² $La_{0.6}Sr_{0.4}Mn_{0.8}Fe_{0.2}O_3$	KB 1270 m² g⁻¹, 15% PTFE	[30]
CB 74 m² g⁻¹, 15%–25% PTFE	3.5–9.9 mg cm⁻² $La_{0.6}Ca_{0.4}CoO_3$	CB 830 m² g⁻¹, 15%–25% PTFE	[31]
70% AB, 30% PTFE	50% $La_{0.6}Ca_{0.4}CoO_3$	35% KB 1300 m² g⁻¹, 15% PTFE	[32]
30%–50% carbon, 50%–70% PTFE	$La_{0.6}Ca_{0.4}CoO_3$	Graphite/graphitized AB, 25% PTFE	[33]
70% CB 47 m² g⁻¹, 30% PTFE	6 mg cm⁻² $LaMnO_3$	85% KB 1270 m² g⁻¹, 15% PTFE	[34]
70% CB 47 m² g⁻¹, 30% PTFE	6 mg cm⁻² $LaMnO_3$	KB 1270 m² g⁻¹, 15% PTFE	[35]
70% CB 47 m² g⁻¹, 30% PTFE	$LaMnO_3$, 20–30 nm	KB 1270 m² g⁻¹, 15% PTFE	[36]
40% AB/active carbon mixture, 60% PTFE	10% $La_{0.6}Sr_{0.4}MnO_3$, 20–50 nm dispersion	50% CB, 40% PTFE vs. Hg/HgO	[37]
70% CB 47 m² g⁻¹, 30% PTFE	$La_{0.4}Ca_{0.6}Mn_{0.9}Fe_{0.1}O_3$, 15 nm	KB 1270 m² g⁻¹, 15% PTFE	[38]
1 mg CMOG powder, 1 mg XC-72 carbon, 50 µL Nafion	$CoMn_2O_4$/graphene (CMOG)	30% CMOG, 50% acetylene black (AB), 20% polytetrafluoroethene	[39]
Carbon paper with 0.5 mg cm⁻² of Pt	$RP\text{-}LaSr_3Fe_3O_{10}$	4 µL ethylene glycol, 10 mg Pd/RP + $LaSr_3Fe_3O_{10}$	[40]
Spinel 70% EB, 30% PTFE	25% $LiMn_2Co_1O_4$	60% CB, 15% PTFE	[41]
70% EB, 30% PTFE	25% $LiMn_{1.4}Co_{0.6}O_4$	60% CB, 15% PTFE	[42]
40% AB, 60% PTFE	$MnCo_2O_4$	Acetylene black	[43]
Catalyst mixtures 40% Vulcan XC-72, 60% PTFE	9.5 mg cm⁻² $La_{0.1}Ca_{0.9}MnO_3$/1.5 mg cm⁻²CoTPP	Ketjan Black & Vulcan	[44]

Sources: Egan, D.R. et al., *J. Power Sources*, 236, 293, 2013; Wang, L. et al., *J. Electrochem. Soc.*, 158(12), A1379, 2011; Takeguchi, T. et al., *J. Am. Chem. Soc.*, 135, 11125, 2013.

Note: All designs incorporated a nickel-mesh current collector.

CB, carbon black; AB, acetylene black.

subject of many investigations because of their modest activity and relatively higher abundance. In particular, Ag exhibits reasonable activity and stability with the price only about 1% that of Pt, rendering it as an attractive ORR catalyst [45]. The electrocatalytic ORR on Ag(111) single crystal surface shows pH-dependent behavior [46]. Furthermore, another study has shown that porous Ag membranes provide electrocatalytic function (with high exchange current density), mechanical support, and a means of current collection in alkaline cathodes [47]. Thus, silver-based catalysts are more adaptive to be used in alkaline electrolytes and are promising cathode materials with good balance between cost and performance.

3.3.3.2 Transition-Metal Oxides

In terms of high energy density applications, manganese oxide [48] (such as MnO, Mn_3O_4, Mn_2O_3, MnO_2, Mn_5O_8, Mn_3O_4, Mn_2O_3, MnO_2, and MnOOH) has received increased attention due to its high chemical stability as well as low cost, low toxicity, and high catalytic activity. The influence of composition on the properties has been scarcely investigated. One trend is that the ORR catalytic activity correlates with the Mn valence [49], but a clear relationship remains to be established. However, it is obvious that the crystallographic structures and morphologies greatly affect the performance of manganese oxides [50]. For MnO_2-based catalysts, the activity follows an order of α->β->γ-MnO_2, which is attributed to a combinative effect of their intrinsic tunnel size and electrical conductivity.

A recent study has shown that the ORR activity of perovskite family correlates primarily to σ^*-orbital (e_g) occupation and secondarily to the extent of transition metal–oxygen covalency, which serve as activity descriptors [51,52]. Also, the surface area–normalized kinetic current densities of four representative oxides, termed specific activity i_s, are plotted as a function of voltage in Figure 3.9, which were reported by Jin Suntivich [51]. The dashed line in Figure 3.9 shows that the intrinsic activity for each catalyst can be assessed by the potential to achieve a given specific ORR current (25 $\mu A\ cm^{-2}_{ox}$). For $LaCu_{0.5}Mn_{0.5}O_3$, $LaMnO_3$, $LaCoO_3$, and $LaNiO_3$, this specific activity current can be reached at potentials of 781($\pm$15), 834($\pm$24), 847($\pm$3), and 908($\pm$8) mV versus RHE, respectively. A higher potential indicates higher electrocatalytic activity for a given oxide. It is interesting to note that oxides

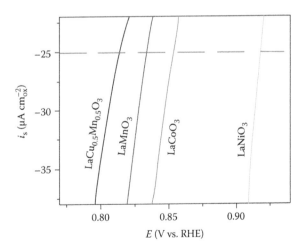

FIGURE 3.9 ORR activity of perovskite transition metal oxide catalysts. Specific activities of $LaCu_{0.5}Mn_{0.5}O_3$, $LaMnO_3$, $LaCoO_3$, and $LaNiO_3$. The potential at 25 $\mu A\ cm^{-2}_{ox}$ is used as a benchmark for comparison (shown as the intersection between the activity and the horizontal gray dashed line). (From Suntivich, J. et al., *Nat. Chem.*, 3, 546, 2011.)

such as $LaMnO_3$ and $LaNiO_3$ have intrinsic ORR activity comparable to the state-of-the-art Pt/C. According to this, the prominent bifunctional electrochemical performances of perovskite composites suggest their promising use in rechargeable metal–air batteries.

3.3.3.3 Other Catalytic Materials

The exclusive use of platinum is not feasible for the future widespread deployment of electrochemical devices based on oxygen electrodes. Development of well-performing noble metal-free cathode catalysts is the ultimate solution to the obstacle of Pt cost and scarcity. One of the carbonaceous [53] materials has attracted extensive attention either as catalyst supports or as metal-free catalysts in electrocatalytic chemistry of oxygen. Also, tremendous effort has been directed to the development of inorganic–organic [54] composite materials as noble metal-free catalysts for oxygen electrochemistry. Both of them are highly valuable cathode catalysts candidate for practical metal–air batteries despite relatively fewer applications to date.

3.3.4 Electrochemical Test Procedures

The performances of the air electrode were evaluated using an electrochemical workstation. The counter electrode was a Pt wire, the reference electrode is Hg/HgO, and the electrolyte was aqueous alkaline solution or neutral saline electrolyte. Electrochemical test procedures were normal ORR test by Rotating disk electrode method and constant discharge method test by electrochemical workstation, which have been introduced in fuel cells and lithium second batteries.

3.3.5 Failure Mode Analysis and Mitigation Strategies

The discharge capacity of most catalysts used so far has shown unsatisfactory results, especially at current densities of substantial use for the power or electric vehicle utilities. One of the major problems associated is how to reduce the overpotential of oxygen reduction.

Electrocatalysts for the oxygen reduction have been searched extensively to date, covering metals, organometals, oxides, sulfides, carbides, and nitrides. Noble metal catalysts like Pt and Ag are fairly active, but less expensive catalysts, if available, are definitely more desirable. In this sense, single and mixed oxides of transition metals have drawn attention of many researchers for the use in alkaline electrolyte. Especially, high expectation has been directed to some kinds of perovskite-type oxides since Meadowcroft [55] first pointed out the potentiality of these oxides to replace Pt catalyst. Perovskites are excellent alternatives to noble metals as low-cost catalysts. Mixed with high-surface-area carbons, perovskites exhibit excellent cathodic oxygen reduction properties in alkaline electrolytes.

3.3.6 Cathode Fabrication

The air electrode is consisted of a microporous gas diffusion layer, a porous nickel foam substrate as a current collector, and a catalyst layer. The gas diffusion layer is prepared by corresponding materials and binder. After mixing materials and binder together and getting a paste, the mixture paste is coated on one side of a porous nickel foam current collector. The catalyst layer of air electrode was prepared by mixing catalyst agents, conductive agents, and binders by weight proportion. The slurry was dried at temperature around 100°C to give a dough-like paste. The catalyst paste was coated on to the other side of the porous nickel foam substrate. The air electrode was finally treated by pressure and high temperature.

3.4   AAB ELECTROLYTES

3.4.1   INTRODUCTION

Like other battery systems, the electrolyte that separates the two electrodes to avoid short circuit and provides OH⁻ for maintaining the electrochemical reactions plays an important part in AAB. The battery performance strongly depends on the choice of electrolyte. Aqueous electrolytes are widely used in Al–air primary battery because of their high ionic conductivity, and the major development effort has focused on two types of aqueous electrolyte, which are alkaline and saline electrolytes. For the alkaline electrolytes, they also have the ability of regulating the reduced oxygen ion into hydroxide anion [3]. However, the Al anode suffers from severe self-discharge and passivation/corrosion in aqueous electrolyte, leading to the low coulombic efficiency and short battery shelf life. Besides, AAB generates heat during both idle and discharge, resulting in high rate of water loss for the aqueous electrolyte and accelerated corrosion rates [56]. It leads to hazardous and runaway conditions and also severely decreases the battery shelf life [57]. To overcome these obstacles, investigation of electrolyte additives and new electrolyte systems are promoted to inhibit the anode corrosion and self-discharge for AAB.

3.4.2   ELECTROLYTE TYPES

3.4.2.1   Aqueous Electrolytes

Aqueous solutions are simple to operate, low in cost, and with less environmental pollution [9]. The aqueous electrolyte widely used in AAB is typically a neutral saline (sodium chloride, potassium chloride, or seawater) solution or alkaline solution (sodium hydroxide or potassium hydroxide) [1].

AABs employing saline electrolytes have been studied mostly for low-power equipment, such as emergency lighting, portable equipment, stationary standby power sources, and marine applications. Compared with alkaline electrolyte, neutral saline electrolytes are less caustic, resulting in a lower corrosion rate and a lower risk to health. The corrosion rate in saline electrolyte is linearly proportional to the current density, and it starts from near zero at zero current [6]. A 12 wt% solution of sodium chloride, which is near the maximum conductivity, is considered as suitable neutral electrolyte. Current densities are limited to 30–50 mA/cm² in consideration of the conductivity of the electrolyte [6]. When operated in seawater, only limited current capability can be obtained because of the lower conductivity of seawater [58].

Alkaline solutions are widely employed because of their high ionic conductivity, the good performance of the air electrode, and the acceptable stability of the base metal electrode [59]. Preferred electrolytes for AABs are 4 mol dm^{-3} NaOH and 7 mol dm^{-3} KOH, which have peak electrolytic conductivity of 0.39 and 0.7 S cm^{-1}, respectively [1]. KOH solution has lower viscosity, higher oxygen solubility and diffusion coefficients. The solubility limit of the reaction product, aluminate, is higher in 7 mol dm^{-3} KOH than in 4 mol dm^{-3} NaOH solution. Therefore, KOH solution shows superiority to NaOH solutions for oxygen reduction [1]. The ionic activity increases with increasing KOH and NaOH concentration. At lower alkaline electrolyte concentrations, the aluminum electrode surface is thicker compared with that at high alkaline electrolyte concentration, which may resist Al^{3+} and OH$^-$ ions transport between the electrode surface and the electrolyte solution [60].

Saline electrolyte has other merits compared with alkaline electrolytes. For example, alkaline electrolytes can absorb carbon dioxide in the air and form the carbonation, which may impede air access and cause mechanical damage to the gas diffusion cathode, and saline electrolytes do not have this risk. However, the anodic dissolution of aluminum in saline solutions results in undesirable hydrated aluminum hydroxides, which causes anode passivation due to the accumulation of aluminum hydroxide on the electrode and increases the water requirement [9]. The high conductivity and Al(OH)$_3$ solubility of alkaline solutions help the system to obtain attractive high power and energy density compared with saline solution.

Combining the alkaline electrolyte with the polymer forms the alkaline polymer electrolyte. The ionic conductivity varies depending on the composition. For the polymer electrolyte working as both ion conductor and separator, it must have high ionic conductivity, good mechanical strength, and high thermal stability. Several kinds of alkaline polymer electrolyte have been developed. PEO/KOH polymer electrolytes exhibited ionic conductivity around 10^{-3} S cm^{-1} at room temperature. PVA/PEO/KOH and PVA/KOH polymer electrolytes have been investigated as electrolyte for Ni-MH and Zn-air batteries [61]. In order to obtain high ionic conductivity and mechanical strength, PVA/PAA/KOH and PAA/KOH solid polymer electrolytes were investigated to use in the AAB system [62,63]. There is some free water in the matrix of PVA/PAA polymer, and the melting temperature of PVA/PAA polymer membrane becomes lower with higher PAA content. The melting temperature was found to be around 190°C–210°C when the PVA/PAA ratio varies from 10:3 to 10:7.5. The PVA/PAA polymer exhibits two-step degradation, the first step is the decomposition of PAA in the temperature range of 320°C–370°C, and the second one is the decomposition of PVA starting at 410°C. Evidently, the PVA/PAA membrane samples are relatively stable in the temperature range of 100°C–300°C. XRD and SEM results show that the translucent polymers are amorphous, exhibiting uniform layer structure morphology with no phase separation. The ionic conductivity of alkaline PVA/PAA polymer membrane electrolyte increases as PAA content is increased at all the temperatures, and higher ionic conductivity is obtained for samples with higher KOH content. However, the higher PAA content lowers the mechanical strength of alkaline PVA/PAA polymer. Therefore, the composition must be optimized to get good balance between enhanced ionic conductivity and reasonable mechanical properties for the solid polymer electrolyte applications.

3.4.2.2 Nonaqueous Electrolytes

Practical Al electrodes operate at significantly lower potentials in aqueous electrolyte because that Al is normally covered with an oxide/hydroxide surface film, resulting in a voltage delay. Al anode experiences a considerable parasitic corrosion reaction in aqueous solutions, which leads to a low Al utilization accompanied by a massive hydrogen gas evolution. For the aqueous electrolyte, progressive consumption of hydroxyl ions occurs at the Al electrode and the electrolyte progressively saturates with aluminate during discharge. When aluminate concentration exceeds the supersaturation, crystalline of aluminum hydroxide precipitates, resulting in loss of the ionic conductivity of the solution.

To resolve these problems, researchers have focused their attention on room temperature ionic liquids (RTILs) as AAB electrolytes since they are extremely nonvolatile, highly stable, and highly conductive. Moreover, Al is not prone to the parasitic hydrogen generation reaction in RTILs. Another attractive advantage is that electrodeposition of Al, which cannot be obtained from aqueous solutions at moderate temperatures, becomes easier in ionic liquid [64]. The evolution of hydrogen occurs before the deposition of aluminum in alkaline electrolyte; therefore, it is essential to explore nonaqueous electrolytes for AAB if rechargeable AABs are to be obtained.

RTILs are low-temperature molten salts, composed mostly of organic ions. The ions often have a delocalized or shielded charge configuration, and this is responsible for the low melting point of the salt. The application of RTILs as AAB electrolytes is a relatively new area of study that is steadily gaining attraction in the electrochemical and engineering communities. RTILs are not currently used in commercial devices due to the high cost and the short history of investigation.

Chloroaluminate ionic liquids exhibit Lewis acid–base chemistry, and chloroacidity is the major determinant in speciation, reactivity, and electrochemistry in chloroaluminate ionic liquids. The composition of the melt determines its chloroacidity. $AlCl_3$/[EMIm]Cl ionic liquids have been investigated as AAB electrolyte [3]. With a $AlCl_3$ mole fraction less than 0.5, these room temperature melts are basic. At $AlCl_3$ mole fractions greater than 0.5, these melts may be regarded as acidic. Room temperature conductivity at the level of 10 mS cm^{-1} is typical of ionic liquids based on EMIm$^+$ cation. Figure 3.10 shows the cyclic voltammetry of $AlCl_3$/[EMIm]Cl ionic liquids. The electrochemical windows of the neutral melt window (4.6 V) is extraordinarily wide compared with the acidic (2.6 V) and basic (3.1 V) ones.

Another kind of nonaqueous electrolyte should be the anhydrous hydroxide conductor solid electrolyte that can be applied in rechargeable AAB. Research in this field has been developed since 2013; therefore, results for rechargeable AAB are quite limited. Hibino et al. have done the investigation on all-solid-state rechargeable AAB applying a hydroxide ion-conducting Sb(v)-doped SnP_2O_7 as electrolyte. This series of compounds have hydroxide ion exchange capability in the bulk of SnP_2O_7 by charge compensation through the partial substitution of Sn^{4+} with Sb^{5+} [65]. $Sn_{0.92}Sb_{0.08}P_2O_7$ exhibited the highest hydroxide ion conductivities above

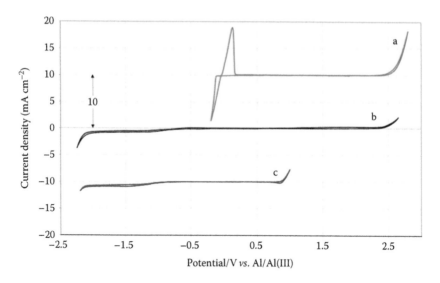

FIGURE 3.10   Cyclic voltammetry at 20 mV s^{-1} of AlCl$_3$/EMImCl melts: (a) acidic, (b) neutral, and (c) basic. (From Revel, R. et al., *J. Power Sources*, 272, 415, 2014.)

0.01 S cm^{-1}, which is much lower than that of aqueous electrolyte but comparable to that of ionic electrolyte. Moreover, Sn$_{0.92}$Sb$_{0.08}$P$_2$O$_7$ shows high tolerance to CO$_2$ in air compared with KOH aqueous alkaline electrolyte.

3.4.3   CORRELATION BETWEEN ELECTROLYTE PROPERTIES AND BATTERY PERFORMANCE

In neutral saline electrolyte, the cell discharge performance is mainly governed by electrolyte conductivity. The investigation results of electrochemical performance characteristic of AAB employing seawater and NaCl electrolyte [58] show that the variation of open-circuit voltage (OCV) is almost of similar trend with the electrolyte conductivity. The cell employing 4 M NaCl showed a operating voltage of 0.7 V and a discharge capacity of 250 mAh, while for the Al–air seawater cell, the operating voltage was around 0.64 V and the discharge capacity was around 150 mAh. Evidently, AAB exhibits high OCV and long discharge duration when the saline electrolytes have high conductivity. Mori studied AAB using 10% NaCl aqueous solution as the electrolyte [66]. The first discharge capacity was 15.91 mAh cm^{-2}; however, the battery voltage fell steeply and the capacity was only 0.11 mAh cm^{-2} at the second discharge process. This is the typical behavior of a basic AAB.

In alkaline electrolyte battery system, the electrolyte concentration has a strong positive effect on the current density and thus the power density. The *i–V* performance and power density of the AAB in NaOH electrolyte are shown in Figure 3.11 [67]. The purity of Al is 97.6 wt% (impurities: O 1.13, Fe 0.68, and Ag 0.59 wt%). The OCV of the battery is around 1.45–1.5 V, which varies little with different concentrations of electrolyte. The short circuit current density ranges from

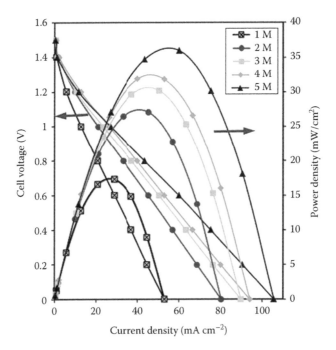

FIGURE 3.11 Performance of the AABs in 1–5 M NaOH electrolyte. (From Wang, L. et al., *Int. J. Hydrogen Energy*, 38(34), 14801, 2013.)

54 to 105 mA cm^{-2}, and the peak power density ranges from 17.5 to 36.2 mW cm^{-2} as the electrolyte concentration increases from 1 to 5 M. Typical discharge profile of AAB in alkaline KOH electrolyte is shown in Figure 3.12 [68]. The operating voltage decreased rapidly in the early discharging stage, which is caused by the battery internal resistance, and then reaches an approximate constant value. The specific energy of AAB employing alkaline electrolyte can be as high as 400 Wh kg^{-1}, making it a candidate for high-power applications [6]. However, in alkaline media the coulombic efficiency of aluminum is low ascribed to the corrosion, and the corrosion current/rate increases with increasing concentration of the alkaline electrolyte [69].

3.4.4 COMPATIBILITY BETWEEN ELECTROLYTE AND SEPARATOR

The separator used in AAB is high-porosity hydrophilic film with low electrochemical resistance and the ability to absorb and retain the aqueous electrolyte to electrically isolate the positive and negative electrodes. The compatibility between electrolyte and separator is quite good in battery employing aqueous electrolyte.

In battery system using all-solid-state electrolyte, the electrolyte can also play a role as separator. Therefore, no actual separator is needed in all-solid-state AAB.

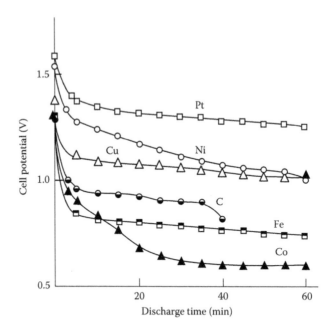

FIGURE 3.12 Discharge profiles for batteries: Al/1 M KOH/metalized graphite air. Graphic symbols in this figure represent different metal catalysts. (From Mukherjee, A. and Basumallick, I.N., *J. Power Sources*, 45(2), 243, 1993.)

3.4.5 Failure Mode Analysis and Mitigation Strategies

In neutral saline electrolyte, the protective oxide film on aluminum metal induces serious polarization problem and causes a decrease in the reversible electrode potential, namely, the AAB voltage is lower than the theoretical value. The time lag before the battery reaches its maximum operating voltage when the circuit is closed [9]. Reaction product aluminum hydroxide accumulates on the electrode, which passivates the anode and further deteriorates the electrochemical performance. Using alkaline electrolyte can resolve these problems to some extent because the protective oxide film can be removed by dissolution in concentrated alkali solutions. Moreover, the solubility of the reaction product in alkaline solution is higher than that in saline electrolyte, which is beneficial for reducing the polarization. However, wasteful corrosion is accelerated in alkaline electrolyte, which is a major barrier to commercialization. In addition, optimizing the electrolyte management can help improve the battery performance. For example, in an exposed battery design that utilizes excess seawater, the discharge performance was greatly extended in comparison with that in the close configurations [58].

In alkaline electrolyte, the high open-circuit corrosion rate of aluminum is the major factor that decreases the efficiency of anode and deteriorates the performance of full cell. From the standpoint of electrolyte, several strategies have been applied to make significant advances in reducing the corrosion of aluminum.

TABLE 3.5
Effect of Solution Phase Inhibitors on Inhibiting the Corrosion Rate of Al in Alkaline Solutions at Open Circuit and under Discharge in Half-Cell Tests

	% Inhibition Efficiency at Open-Circuit	% Discharge Efficiency under Galvanostatic Discharge		
		% ϕ_{dis} at j in mA cm^{-2}		
Inhibitor (mol dm^{-3})	% ϕ_{inh}	100	200	400
Pure Al in uninhibited solution [70]		70	89	
Stannate				
10^{-4} Na$_2$SnO$_3$ [70]	−94			
10^{-3} Na$_2$SnO$_3$ [70]	24	79	95	
10^{-2} Na$_2$SnO$_3$ [70]	67	63	77	
Indium hydroxide				
10^{-4} In(OH)$_3$ [70]	97		50	
10^{-3} In(OH)$_3$ [70]	−178			
10^{-2} In(OH)$_3$ [70]	−195	30	49	
10^{-3} K$_2$MnO$_4$ [70]	32	91	99	
10^{-3} K$_2$MnO$_4$+10^{-3} In(OH)$_3$ [70]	−5	70	87	96
10^{-3} K$_2$MnO$_4$+10^{-2} In(OH)$_3$ [70]	26	69	87	
10^{-2} Na$_2$SnO$_3$+10^{-3} In(OH)$_3$ [70]	80	76	88	96
10^{-2} Na$_2$SnO$_3$+10^{-2} In(OH)$_3$ [70]	73	79	89	96
0.2 M ZnO [71]	98			
Cationic surfactants				
1.8×10^{-4} CTAB [72]	14			
0.5×10^{-3} CTAC [73]	40			
2.976 g L^{-1} lupine [72]	63			

Sources: Macdonald, D.D. and English, C., *J. Appl. Electrochem.*, 20(3), 405, 1990; Wang, X.Y. et al., *Mater. Corr.*, 61, 1, 2010; Abdel-Gaber, A.M. et al., *Mater. Chem. Phys.*, 124, 773, 2010; Al-Rawashdeh, N.A.F. and Maayta, A.K., *Anti-Corr. Meth. Mater.*, 52, 160, 2005.

Modifying the electrolyte by adding inhibitors or additives can make the electrolyte less corrosive. Table 3.5 shows the effect of solution phase inhibitors on inhibiting the corrosion rate of Al in alkaline solutions at open circuit and under discharge in half-cell tests [70–73]. In Table 3.5, 1×10^{-3} mol dm^{-3} Na$_2$SnO$_3$ was found to be the ideal stannate, with the discharge efficiency as high as 95%. The outstanding performance is ascribed to the inhibiting effect on water reduction by plating out tin onto cathodic sites on the surface of aluminum [70]. Of the single-component inhibitor systems, electrolyte containing K$_2$MnO$_4$ was the most effective, with the coulombic efficiency of the aluminum anode exceeding 90%. In fact, oxide/hydroxide of gallium, indium, calcium, and zinc as well as stannates and citrates have been found to be effective in inhibiting corrosion and/or boosting the electrode potential in the alkaline electrolyte AAB system [9].

TABLE 3.6

Capacity and Energy Values at Different Current Densities

Discharge Current Density (mA cm^{-2})	Median Voltage (V)	Capacity (Ah)	Energy (Wh)	Energy (per kg of Aluminum) (Wh kg^{-1})	Ratio of Al Used (%)	Energy (per kg of Reacted Aluminum) (Wh kg^{-1})
−100	0.67	0.125	0.084	86.2	4.3	1995
−300	0.55	0.085	0.047	47.4	2.9	1638
−600	0.45	0.050	0.023	23.6	1.8	1340

Source: Revel, R. et al., *J. Power Sources*, 272, 415, 2014.

Due to the nature of the alkaline solvent's reactivity with aluminum, the corrosion reaction will occur even if inhibitor/additive is added in alkaline electrolyte. Therefore, researchers try to figure out this problem in another way, which is changing the aqueous electrolyte to nonaqueous one, and designing an all-solid-state solvent system that can prevent corrosion. AAB employing EMImCl/AlCl$_3$ RTIL as electrolyte has an exceptional reduced self-discharge rate [3]. The corresponding capacity and energy values at different current densities are exhibited in Table 3.6.

Furthermore, all-solid-state AAB is a promising rechargeable system. Two kinds of all-solid-state battery system were developed. One kind is battery applying polymer alkaline gel as electrolyte whose ionic conductivity is primarily determined by the contents of alkaline and the polymer agent. Zhang et al. have reported an all-solid-state battery using PAA-based alkaline gel electrolyte, which have a high ionic conductivity of 0.46 S cm^{-1} [62]. Figure 3.13 displays the electrochemical performance of this battery system. High area densities of 29.2 mA h cm^{-2} and 30.8 mWh cm^{-2} and high power density of 91.13 mW cm^{-2} were obtained. Although the polymer alkaline gel electrolyte has been indicated to show excellent performance in anodic protection, Al corrosion cannot be ignored especially during low current discharging or rest [62]. The other kind is battery using anhydrous hydroxide ion conductor as electrolyte. For example, by applying Sn$_{0.92}$Sb$_{0.08}$P$_2$O$_7$ hydroxide ion conductor [65], the electrodeposition of aluminum could be accomplished without the hydrogen generation. Besides, this kind of electrolyte shows high tolerance to CO$_2$ in air, which is an additional advantage over alkaline electrolytes. The battery generated OCVs between 1.53 and 1.67 V, which are near the values observed for batteries fabricated with KOH-based electrolytes.

Though using solid-state electrolyte could be an effective strategy to prevent corrosion, the specific energy of all-solid-state system is far from satisfactory compared with that applying aqueous electrolyte, owing to the low conductivity. However, all-solid-state AAB may show promise in the future as the technique is developed.

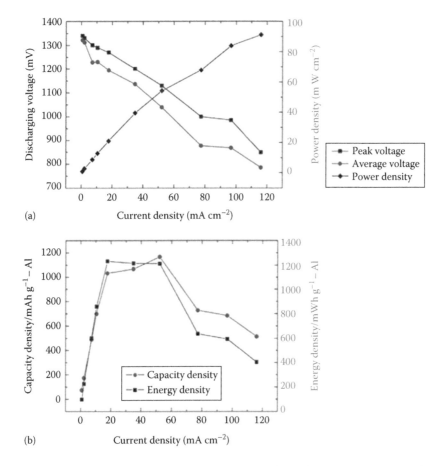

FIGURE 3.13 (a) Discharging voltage and power density profiles versus current density; (b) capacity and energy density profiles during constant current discharge. (From Zhang, Z. et al., *J. Power Sources*, 251, 470, 2014.)

3.5 AAB SEPARATORS

3.5.1 INTRODUCTION

A separator is an important component in AAB to prevent internal short circuits. It should keep chemical and electrochemical stability in the system, good interfacial compatibility with electrolytes, in combination with good ionic conductivity. The separator typically consists of nonwoven laminated polyolefine membranes or glass fibers, and is modified to improve its wettability so that it accommodates a sufficient amount of electrolytes [59]. All-solid-state AABs based on polymer or ceramic electrolyte membranes have also been used as an electrolyte in order to trap the solution and minimize any electrolyte leakage.

3.5.2   Separator Types and Their Physicochemical Properties

Typical polyolefine membranes are Celgard® (manufactured by Celgard LLC, Charlotte, NC). They are polyethylene (PE), polypropylene (PP), and the composite of the two polyolefines. Commercial separators possess a pore diameter of 0.03–0.1 μm and a porosity of 30%–50%. The melting points of PE and PP are 135°C and 165°C, respectively.

Poly(vinyl alcohol) (PVA)- and poly(acrylic acid) (PAA)-based solid polymer electrolyte membranes with high ionic conductivity, good thermal, and mechanical properties were also used in AAB batteries. The highest room temperature value in ionic conductivity for the alkaline solid PVA/PAA polymer electrolyte membranes was 0.301 S cm^{-1} when the weight ratio of PVA:PAA is 10:7.5. Uniform morphology was achieved in the membrane structure. There was no apparent phase separation and the PVA/PAA polymer membrane samples appeared translucent under illumination [63].

Yuxin Zuo et al. prepared the polyacrylic acid (PAA)-based alkaline gel electrolyte used in all-solid-state AAB to prevent leakage. An alkaline solution with pH 14.7 and monomers with a cross-linker were gelled by a polymerization initiator. The optimal gel electrolyte exhibits an ionic conductivity of 460 mS cm^{-1} at 25°C, which is close to that of aqueous electrolytes [62].

A solid electrolyte based on $Al_2(WO_4)_3$ and Sb(V)-doped SnP_2O_7 was also investigated for secondary AAB in recent years.

3.5.3   Correlation between Separator Properties and
Battery Performance

The cell power and energy densities of AAB based on polyolefine separators are up to 175 W kg^{-1} and 400 Wh kg^{-1}, respectively [6].

Solid polymer electrolyte membranes are key components in the development of an electrochemical or battery system. Some polymers that exhibit hydrophilic characteristics have received much attention for electrolyte membranes in recent years. The first theoretical study of solid polymer electrolyte was carried out by Wright et al. [74,75] and was devoted to the ionic nature for the conducting behavior. Armand et al. [76] described the potential use of solid polymer electrolytes in secondary batteries. Later, Fauvarque et al. reported that alkaline PE-potassium hydroxide (KOH) polymer electrolytes exhibited room temperature ionic conductivity around 10^{-3} S cm^{-1} [77,78].

The PAA-based alkaline gel electrolyte is used in all-solid-state AAB to prevent leakage. The optimal gel electrolyte exhibits an ionic conductivity of 460 mS cm^{-1} at 25°C, which is close to that of aqueous electrolytes. The AAB peak capacity and energy density considering only Al can reach 1166 mAh g^{-1} and 1230 mWh g^{-1}, respectively, during constant current discharge. The battery prototype also exhibits a high power density of 91.13 mW cm^{-2}. If the battery is a laminated structure, area densities of 29.2 mAh cm^{-2} and 30.8 mWh cm^{-2} are presented to appraise the performance of the whole cell [62].

A KOH solution gelled with hydrophonics gel (HPG) has been used as an electrolyte for AAB in order to trap the solution and minimize any electrolyte leakage [79].

The energy density was very low at 183 W h kg^{-1}-Al due to the extremely high resistance of the gel restricting the discharge current density to below 10 mA cm^{-2}, the use of a 95 wt% pure anode, and anode passivation by a layer of Al(OH)$_3$.

The solid-state PVA/PAA polymer electrolyte membranes were prepared by a solution casting method from PVA polymer, acrylic acid monomer, a cross-linker, and KOH after polymerized by an initiator. The resulting homogeneous polymer solution was then used for solution casting to form PVA/PAA polymer membranes. The PVA/PAA polymer membranes were further immersed in 32 wt% KOH solution to obtain the gel separators. The cell's discharge performance was extremely poor with a peak power density of 1.2 mW cm^{-2} at a low current density of 1.2 mA cm^{-2} due to passivation of the aluminum surface by the discharge product [63].

The conductivity of gel electrolytes would need to be improved for its viable usage. Gel and solid polymer electrolytes have low solubility for Al(OH)$_4$ ions meaning that during anodic discharge the Al(OH)$_3$ cannot dissolve into the electrolyte to form aluminate. The conductivity of these electrolytes is also insufficient to replenish the OH$^-$ ions consumed at the electrode surface, resulting in passivation of the aluminum.

A ceramic aluminum ion conductor Al$_2$(WO$_4$)$_3$ can prevent anode corrosion due to direct contact with the alkaline electrolyte while retaining aluminum ion conduction. R. Mori has combined both approaches by placing Al$_2$(WO$_4$)$_3$ as an aluminum ion conductor, both on the anode and air cathode sides, and succeeded in preparing AAB with stable cell properties [80,81]. Sb(V)-doped SnP$_2$O$_7$ with hydroxide ion exchange capability was also investigated as a solid electrolyte for rechargeable AAB [65]. This battery generated an OCV of 1.6 V with a discharge capacity of 800 mAh g$^{-1}_{electrode}$.

3.5.4 FAILURE MODE ANALYSIS AND MITIGATION STRATEGIES

For secondary AAB, Al dendrite growth during long-term cycling often breaks through separators. This leads to short circuits and sometimes even catastrophic failure. The mitigation strategy is to reinforce the strength of separators by establishing a strong mechanical barrier to suppress the dendrite growth. Some researchers have attempted to trap electrolyte solution into a hydroponic gelling agent or polymer membrane to minimize electrolyte leakage. These solid-state electrolytes based on polymers or ceramic with good ionic conductivity can be a promising candidate for the separator of AAB. Another approach is adding additives into the electrolytes to block dendrite growth, as lithium–air batteries, but there are few studies about it.

3.6 AAB CURRENT COLLECTORS

3.6.1 INTRODUCTION

The current collector of the air electrode should have high electrical conductivity, gas permeability to allow oxygen diffusion, and stability with respect to the oxidation power of oxygen. A nickel mesh is generally used for this purpose [59]. What's more, a nickel, stainless steel, titanium slice, or mesh and glassy carbon rod were also prelaminated onto the carbon electrode as a current collector.

A copper, nickel mesh, or copper disk was pressed into the aluminum foil and extended to the outside of the cell as the anode current collector. Then AAB were assembled by stacking the cathode-tape disk, one separator disk soaked in the electrolytes, and one aluminum disk [82].

3.6.2   Current Collector Types and Their Physicochemical Properties

The current collectors of AAB should keep chemical and electrochemical stability and have good conductivity in the electrolytes in open-circuit or charge-discharge condition.

The composition of the cathode current collector electrode includes nickel, stainless steel, and titanium. They can be mesh or slice. For the anode current collector, a copper, nickel mesh, or copper disk is mostly used.

3.6.3   Correlation between Current Collector Properties and Battery Performance

The conductivity and weight of electrochemically and chemically stable current collectors can influence the performance of AAB. For high power density of AAB, current collectors with good conductivity are in favor. The smaller the weight of current collectors is, the higher the specific energy AAB can output. During large current discharge, the conductivity and cross-sectional area of current collectors should be enough to support the large current. Otherwise, a huge amount of heat is generated, which results in a large resistance impairing the performance of AAB.

3.6.4   Failure Mode Analysis and Mitigation Strategies

The current collectors of AAB sometimes can be corrupted by the electrolytes during working process. The sealing of AAB is necessary to keep the batteries in good condition, and effective heat management method can also benefit for AAB operation during both idle and discharge periods.

Besides, it may come into being serious problems if AABs get improper management during the operation, such as overcharging, overdischarging, and imbalance of voltage among the cells. So in AAB program, designing reliable controlling system is essential on the working platform.

3.7   AAB MANUFACTURE

3.7.1   Introduction

The world has a lot of interest, and motivation comes from fossil energy–based electricity to electricity generated from renewable energy, such as solar or wind power generation. Nowadays, large solar or wind power generation is feasible. For meeting the continuous energy needs and effectively balancing the cycle of these energy properties, the development of new energy storage system is very important. Several groups have developed a potential of these characteristics based on the ionic liquid electrolytes for rechargeable AAB. The performance of the aluminum battery application is summarized: "aluminum is a very attractive anode material for energy storage and conversion." [83]

The high energy density and theoretical capacity belonging to aluminum is a huge advantage. Some other advantages of AAB with the production of a solid in liquid electrolyte solution has been discussed for vehicle propulsion, but with the formation of the oxide film, the high corrosion rates of aluminum, parasitic hydrogen evolution, and decrease in reversible electrode potential are serious problems and eager to be settled down to realize the practical application. Therefore, the AAB manufacture is the indispensable factor, which should be considered for the practical applications.

The United States, Canada, Israel, and other countries in the development of AAB as a power supply improve the utilization of aluminum, reducing the cost of air electrode catalyst and increasing the life cycle. In the late 1980s, the former Yugoslavia Belgrade Institute of metallurgy and the U.S. electric technology research company have developed neutral and alkaline aluminum air battery by adding an aluminum electrode for 1600 km of operation of the electric vehicle; in the 1990s, Al launched the energy density for 220 Wh kg^{-1} AAB; more than 4000 electric cars equipped with that have trial run; the United States launched the energy density of AAB, 300 Wh kg^{-1} in 1994, and a breakthrough in the integrated technology. The battery capacity can reach above 5000 Ah; America Laurence Livermore National Laboratory in the United States funded by the Department of energy replaced the internal combustion engine (ICE) with the metal air battery. The Dow Chemical company associating with Voltek company successfully improved battery efficiency from 65% to more than 90%; in 2014, Israel Phinergy company and Alcoa Corp have made progress in the miniaturization of AAB, and the two companies are cooperating to develop AAB with the density energy 300 Wh kg^{-1}, so that 100 kg AAB Citroen C1 electric car can reach 1600 km.

3.7.2   ANODE MANUFACTURE

Aluminum is a good anode material, with standard electrode potential in neutral electrolyte −1.65 V (vs. SCE) and in the strong alkaline electrolyte 2.35 V (vs. SCE), but the electrode potential of the aluminum anode in the strong alkaline battery can be moved to −1.5 to 2.0 V and will be changed to about −1.2 V at the discharge current density of 100 mA cm^{-2}. This situation is caused by the following several reasons: First, the passivation film on the surface of the aluminum surface can cause the electrochemical activity of the aluminum to be suppressed. Second, aluminum as a two-property-metal element, in the strong alkaline electrolyte environment, has the reaction of a serious hydrogen evolution corrosion, resulting in electrode potential positive shift and decrease of the battery current efficiency. The corrosion reaction product of aluminum hydroxide colloid and the electrolyte conductivity is decreased and the corrosion will not stop even at the rest state. So anode manufacturing is really a big problem that needs to be solved.

To solve that, two methods are used to improve the heat treatment process of aluminum alloy. Industrial grade aluminum (99.0%) contains impurities, such as iron (0.5%), silicon, copper, manganese, magnesium, and zinc. It will result in the corrosion of hydrogen evolution at the interface of aluminum intensifies; and the existence of iron can result in increasing exponentially for the electrochemical corrosion. Some elements can be added into the aluminum and not only improve the chemical activity but also enhance the resistance to corrosion of the alloy composition, such as gallium indium, tin, zinc, bismuth, cadmium, lead, and so on; the destruction of the passivation film on alumina and aluminum electrode could meet the requirements of large current discharge. Shu et al. [84] propose that aluminum alloys with lead, mercury, cadmium, tin, indium, and other elements have a higher hydrogen evolution overpotential and can improve the hydrogen evolution potential and inhibit the corrosion of aluminum, meanwhile improving the utilization rate of the anode. Manganese offsets the adverse effects of iron in the alloy because manganese can combine with the iron to form $(Fe,Mn)Al_6$ similar to the matrix aluminum chemical activity and does not exacerbate the matrix corrosion, and mercury, lead, and thallium are toxic. In summary, gallium indium and magnesium are the most important basic elements; for the activation and corrosion protection of aluminum, and lead, bismuth, tin, zinc, cadmium, and manganese can be added appropriately in order to enhance the electrochemical performance.

The current efficiency and corrosion morphology of the aluminum electrode depend on the microstructure. In addition to the influence of alloying, microstructure is also influenced by heat treatment process. Heat treatment is mainly by changing the distribution of trace elements in the aluminum alloy and the microstructure of the alloy surface, which affects the properties of the alloy. The optimum heat treatment process can be found by the appropriate orthogonal experiment. Xiaofeng Liu summarized the effects of quenching and annealing heat treatment process of aluminum and chemical properties: after annealing, the aluminum alloy material is activated, the open-circuit electrode potential is about 300 mV, and the working electrode potential is negatively shifted to 10 mV; after quenching, the working potential of the aluminum alloy material is negatively shifted from 20 to 40 mV. Heat treatment technology can reduce the vacancy dislocation and other structural defects in the crystal structure of the aluminum alloy, so that the uniformity of the alloy can be improved and can reduce the corrosion of the anode.

3.7.3 Cathode Manufacture

Air electrode is the key to the component of AAB and is also the core of related research. The air electrode is essentially an oxygen electrode, and the working principle is the theory of gas diffusion electrode. They should form as many effective gas–solid–liquid–phase electrochemical active sites as possible in the electrode. The air electrode is composed of three layers: the porous catalytic layer, the metal mesh current collector, and the waterproof layer. The porous catalytic layer is the main site of oxygen reduction, where the diffusion of oxygen reduction catalyst and the formation of a three-phase interface at the junction of the thin layer electrolyte; the conductive current collector has the main function of conduction and mechanical support; the waterproof breathable layer has the structure that is porous and hydrophobic,

TABLE 3.7

Preparation Method of the Air Electrode

Positive Electrode Composition	Raw Material	Preparation Method
Porous catalytic layer	PTFE, catalyst, catalyst carrier	Rolling, spread coating, sputtering, screen printing, etc.
Conducting current collector	Foamed nickel, nickel net, etc.	Can be suppressed with the other two into the electrode
Waterproof breathable layer	PTFE, pore-forming agent, activated carbon, anhydrous alcohol, etc.	Rolling, spread coating, etc.

which can provide the required gas to the catalytic layer and prevent the electrolyte. The structure and manufacture of the air electrode is shown in Table 3.7.

Study on the mechanism of oxygen reduction shows that we should develop and manufacture a three-phase gas diffusion electrode and the catalyst (with high conductivity, stable chemical properties, and high catalytic efficiency).

Oxygen reduction catalysts are mainly four kinds of materials: platinum and other precious metals, perovskite-type composite oxide, manganese oxide, and metal chelate (transition metal chelate). Quaino et al. [85] believe that a noble metal catalyst has a high catalytic activity for the reduction of oxygen: oxygen can be achieved by the four-electron reduction process, but the price of the noble metal catalyst which easy to be poisoned is expensive; the molecular formula of perovskite-type oxides is $ABO3$, which is currently the hot spot of the research on oxygen reduction catalyst because of the more choice of elements, high matching, and good catalytic effect; the A-site cations are generally the ions of alkali metals (K, Na, and Rb), alkaline earth metals (Ca, Ba, and Sr) or rare earth metals (La, Cs, and Pr); B cations are generally Fe, Ti, or Co ions; manganese oxide catalysts mainly include Mn_2O_3, MnO_2, and so on, which have good stability but not outstanding catalytic activity. Metal chelate catalyst mainly has the transition metal chelate metal Ye Linlei and phthalocyanine. In the case of cobalt or iron porphyrin, its face structure is conducive to the destruction of oxygen O=O bonds, so it has a very high catalytic activity. But it is not stable or dissolved in acidic or alkaline solution so that the application is limited.

3.7.4 Electrolyte Preparation

The electrolytes of AAB are typically the neutral salt solution or strong alkaline solution. When using the neutral electrolyte, self-corrosion of the anode is not obvious, but aluminum anodic passivation of the surface is serious. Therefore, this type of battery can only be used for powering output device at low current density. While using a strong alkaline electrolyte, the passivation of aluminum decreases and the reaction product can absorb a certain amount of alkali hydroxide. However, aluminum is the amphoteric metal, and a strong hydrogen evolution corrosion will reduce the output power in the strong alkaline environment.

The following two aspects can solve the earlier problems: first, regular replacement of electrolyte, or using the form of electrolyte circulation, and second, additives for the activation of aluminum anode surface and the inhibition of hydrogen evolution in the electrolyte. It can be concluded that existing additives are mainly divided into three kinds [1,9,59]: (1) inorganic ion, such as Cl^-, F^-, SO_4^{2-}, SnO_3^{2-}, Bi^{3+}, In^{3+}, Ga^{3+}, and so on, in which Cl^- and F^- both activate aluminum, but Cl^- can aggravate pitting corrosion of aluminum; (2) organic active agent, such as EDTA and ethanol; and (3) compound additive. The improvement mechanism of compound additives on the properties of electrolytes is relatively complex, and various influencing factors interact with each other. So further research is still needed. Weihua Wan et al. [86] made K_2MnO_4 and citrate compound as the electrode additive, which increases the utilization of the battery aluminum anode from 28.5% to 81.0% and decreases the current density of self-corrosion from 123.8 to 17.87 mA cm^{-2}.

3.7.5 CURRENT COLLECTORS, SEAL, AND SEAL CASE MANUFACTURE

The current collector of the air electrode should have high electrical conductivity, gas permeability to allow oxygen diffusion, and stability with respect to the oxidation power of oxygen. A nickel mesh is generally used for this purpose. The catalyst is sometimes loaded on the current collector [59].

In some experiments, the carbon slurry for the gas diffusion layer was prepared with a mixture of acetylene black (AB50) and PTFE (Teflon-30 suspension, Du Pont) as a binder. The slurry was coated onto the Ni-foam current collector plate, then pressed at a pressure of 10 MPa. The active layer of the air electrode was then prepared by spraying a mixture comprising PTFE, $KMnO_4$, VulcanXC-72, and an appropriate amount of isopropyl alchohol (IPA) onto the gas diffusion layer. The air electrode with both the gas diffusion layer and the active layer on the Ni-foam matrix was finally sintered [63].

The laboratory-made AAB comprised the Al anode, a spacer, and an air cathode. The Al mixture was carefully pressed on the surface of a copper collector to form Al anode for investigating the discharge performance [87].

AAB as a kind of new high-performance energy can be widely used in standby power supply, portable power supply, electric power, water power, and other military civilian areas due to the different application occasions and environment, so its design mode is different, which can be designed into an open type and can also be designed into a closed type. On land, power supply is generally designed to consider the use of oxygen in the air, with the blower providing the required oxygen and solving the problem of heat dissipation. And underwater vehicle power would generally use the compressed oxygen, liquid oxygen, and hydrogen peroxide, but also consider the heat dissipation, hydrogen removing.

3.7.6 BATTERY ASSEMBLING

Depending on the use of the electrolyte, the assembly process of AAB is different.

3.7.6.1 Liquid Electrolyte Aluminum–Air Battery [88]

Three-electrode cell preparation: Three-electrode cell configuration was used, with an Al foil or air electrode serving as the working electrode. Platinum foil was used as a counter electrode, and Fc/Fc$^+$ (ferrocene/ferrocenium ion) gel based was utilized as a reference electrode. The experiments were carried out in a PP electrochemical cell with an exposed surface area with nickel metal connectors. The Al electrode was cleaned in ethanol and acetone and then air-dried. No pretreatment was applied to the air electrode.

Two-electrode cell preparation: In full battery studies (two-electrode configuration), a similar cell structure (to the three-electrode cell) was used but without the introduction of a reference electrode. Al foil and air cathode were utilized as the electrodes. The cells were held at OCP in all the electrochemical experiments for 4 hours prior to initiating the discharge process. This was performed in order to allow a proper wetting of the porous carbon (at the air cathode) with the electrolyte.

3.7.6.2 Gel Electrolyte–Based Aluminum–Air Batteries [79]

This cell was prepared by using a cylindrical plastic casing cell (Figure 3.14). The air cathode sheet was cut into a circular shape with a small portion left for the positive terminal. The circular-shaped Al plate with a small portion (left for the negative terminal) was used as the anode.

3.7.6.3 All-Solid-State AABs [62]

The alkaline gel electrolyte film was sandwiched between an Al anode mesh and an air cathode plate, as shown in Figure 3.15. The laminar structure was easy to fabricate and refuel by replacing the Al mesh (Figure 3.15).

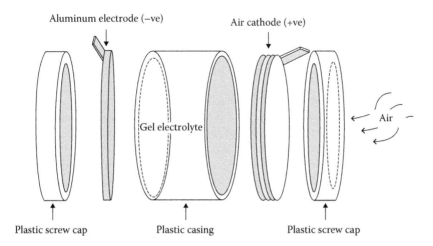

FIGURE 3.14 Schematic diagram of the AAB assembled from its components. (From Mohamad, A.A., *Corros. Sci.*, 50(12), 3475, 2008.)

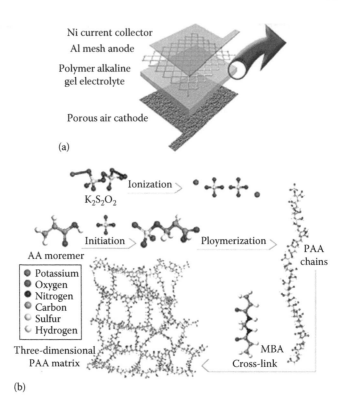

FIGURE 3.15 (a) Schematic of all-solid-state AAB with a laminar structure. (b) Illustration of the polymerization process from AA monomers to a PAA matrix. Particles represent different atoms distinguished by color, white for hydrogen, gray for carbon, red for oxygen, yellow for sulfur, blue for nitrogen, and purple for potassium, respectively. (From Zhang, Z. et al., *J. Power Sources*, 251, 470, 2014.)

3.7.7 Battery Maintenance

AAB is a nonpolluting, long-lasting, stable, and reliable power supply, but also a very environment-friendly battery. AAB has many advantages, such as large specific energy, light weight, no toxicity and risk, and so on. In addition, another advantage of AAB is maintenance convenience. According to the technical scheme, users do not need to charge their batteries as long as (every 1–2 months) they inject water or replace the useless electrolyte to support the chemical reaction, assigning a technical staff carry out the maintenance every year. After a long time, the anode electrode is consumed and only needs to replace the anode aluminum module.

3.7.8 Battery Tests and Performance Diagnosis

When the parts are assembled into a required battery, we need to test the performance of the batteries. The discharge performance of the AABs was studied by constant current discharge testing at different current densities. The Land test system was used. The weight of the anodes was measured both before and after discharge. The anode utilization (η), capacity density, and energy density were calculated using the following formulas [89]:

$$\eta = 100It/(\Delta mF/9) \tag{3.9}$$

$$\text{Capacity density} = Ih/\Delta m \tag{3.10}$$

$$\text{Energy density} = UIh/\Delta m \tag{3.11}$$

where
  η is the anode utilization, %
  I is the current, A
  Δm is the weight loss, g
  F is the Faraday constant
  t is the time, s
  U is the average voltage, V
  h is the discharge time, h

Conductivity measurements were performed at room temperature by impedance spectroscopy with a commercial sensor with internal digitated electrodes on a polyimide support. Electrolyte electrochemical windows were measured using a cycling voltammetry technique with a tungsten working electrode rotating at 250 turn min^{-1} and a scan rate of 20 mV s^{-1}. Corrosion rates were measured applying a linear polarization on an aluminum wire as working electrode and a scan rate of 1 mV s^{-1} for a potential range of ±50 mV around the OCV [3].

3.7.9 BATTERY SAFETY

In high-power output, the heat output system must be equipped with the heat output system. Another security problem for the AAB is the leak. In practical application, how to ensure the battery does not leak is also a serious problem to be solved. Battery in the actual use of the process is needed to avoid the bumps of the battery and prevent fluctuations in output current. There are potential safety hazards in the presence of localized electrochemical reactions in the presence of impurities in aluminum.

3.8 AAB TECHNOLOGY CHALLENGES AND PERSPECTIVES

As a coin, everything has two sides: challenges and perspectives. There are a number of technical problems or challenges that need to be solved to make AAB suitable for practical applications. As mentioned in former sections, anodes made of pure aluminum are severely corroded by the electrolytes (especially in the aqueous alkaline electrolytes), so the aluminum is usually alloyed with tin or other elements. Additionally, the hydrated alumina created during the reaction process forms a gel-like substance at the anode and reduces the electricity output, which is a big problem that should be considered in the development work on AAB. Therein, additives can be one of the solutions that make the alumina to be formed the powder rather than the gel. The air cathode is made from polymers and catalyst layer, which can facilitate oxygen diffusion and ensure the normal operation of the system. In view of the importance of the cathode, further development work on air cathodes is really necessary, including air-permeable membranes, carbon, catalysts, and so on. Additionally, the cell design and maintenance are also the factors needed for consideration.

However, the promising perspectives still exist for AAB. Aluminum as the "fuel" for vehicles has been studied by Yang and Knickle [5]. They have analyzed it from many aspects, that is, driving ranges, cost, and total fuel efficiency. The AAB system can generate enough and power for driving ranges and acceleration similar to gasoline powered car. The cost of aluminum as an anode can be as low as $1.1 kg^{-1} as long as the reaction product is recycled. The total fuel efficiency during the cycle process in AAB electric vehicles can be 20% comparable to that of ICE vehicles (13%). The designed battery energy density is projected as 2000 Wh kg^{-1}. The cost of a battery system can be $29 kW^{-1}. AAB EVs life-cycle analysis was conducted and the AAB EVs can be projected to have a travel range comparable to ICEs. From this analysis, AAB EVs are the most promising candidates for replacing ICEs. Besides that, AAB with limited shelf life is no longer the case with modern designs. So these batteries can be used as reserve batteries in some telephone exchanges as a backup power source. AAB also could be used to power laptop computer, cell phones, other consumer electronics, and so on.

3.9 SUMMARY

AABs have great potentials for using as low-cost, high-performance energy-storage systems, so AAB has attracted great much attention from our researchers. Due to the existence of the overpotentials in the two electrodes and the consumption of water, the practical operation voltage, power, and energy are lower than that of the theoretical. However, though, the performance of AAB still can surpass that of most battery systems. This is the reason that we concretely introduce the AAB in the Chapter 3.

We have introduced the air cathode, aluminum anode, the AAB electrolytes, separators, current collectors, and something about AAB manufacture. Typically, the composite air electrodes are composed of the high-surface-area carbon materials, the catalysts, Teflon binder, air-permeable membranes, and metal mesh current collector. Further development work on air cathodes is really necessary. The electrolytes used in AABs include aqueous alkaline solution, aqueous neutral saline, ionic liquid, solid-state electrolyte, and so on. Some electrolyte additives are also needed to improve the performance of AAB. Anodes made of pure aluminum are severely corroded by the electrolytes (especially in the aqueous alkaline electrolytes), so the aluminum is usually alloyed with tin or other elements. Additionally, the cell design and maintenance are also the aspects mentioned in the last sections of the chapter.

In summary, the AAB with promising perspectives for applications in many fields can become the ideal power in the twenty-first century, but the road for solving the problems is tough and long.

REFERENCES

1. D.R. Egan et al., Developments in electrode materials and electrolytes for aluminium–air batteries, *Journal of Power Sources*, 236 (2013), 293.
2. C. Daniel and J.O. Besenhard, Metal-air batteries, in *Handbook of Battery Materials*. Wiley-VCH Verlag GmbH & Co. KGaA, Weiheim, Germany (2011) p. 759.
3. R. Revel, T. Audichon, and S. Gonzalez, Non-aqueous aluminium–air battery based on ionic liquid electrolyte, *Journal of Power Sources*, 272 (2014), 415.
4. Cleantech Canada Staff, http://www.canadianmanufacturing.com/environment-and-safety/alcoa-phinergy-developing-aluminum-air-ev-battery-138049/, June 2, 2014.
5. S. Yang and H. Knickle, Design and analysis of aluminum/air battery system for electric vehicles, *Journal of Power Sources*, 112(1) (2002), 162.
6. D. Linden and T.B. Reddy, Metal/air batteries, in *Handbook of Batteries*. McGraw-Hill, New York (2001).
7. M. Nestoridi, Al air batteries: The concept—Electrochemistry, in *The Study of Aluminium Anodes for High Power Density Al-Air Batteries with Brine Electrolytes*. PhD thesis, University of Southampton, Southampton, U.K. (2009) p. 3.
8. S.-I. Pyun and S.-M. Moon, Corrosion mechanism of pure aluminium in aqueous alkaline solution, *Journal of Solid State Electrochemistry*, 4(5) (2000), 267.
9. Q. Li and N.J. Bjerrum, Aluminum as anode for energy storage and conversion: A review, *Journal of Power Sources*, 110(1) (2002), 1.
10. H.Z. Wang et al., A review on hydrogen production using aluminum and aluminum alloys, *Renewable and Sustainable Energy Reviews*, 13(4) (2009), 845.
11. J.A. Hunter, Aluminium batteries, EP326338A2 (1989).
12. C.D. Tuck, Aluminum air batteries, in *Modern Battery Technology*. Ellis Horwood, West Sussex, England (1991) pp. 489–490.
13. J.A. Hunter et al., Aluminium batteries, US5004654 A (1991).
14. K.R. Van Horn, Aluminum and alloy, in *Aluminum: Properties, Physical Metallurgy and Phase Diagrams*. American Society for Metals, Ohio (1967), pp. 49–54.
15. K.-K. Lee and K.-B. Kim, Electrochemical impedance characteristics of pure Al and Al-Sn alloys in NaOH solution, *Corrosion Science*, 43(3) (2001), 561.
16. J.A. Hunter, The anodic behaviour of aluminium alloys in alkaline electrolytes. PhD thesis, University of Oxford, Oxford, England (1989).
17. C. Zhang and S.M. Park, The anodic oxidation of nickel in alkaline media studied by spectroelectrochemical techniques, *Journal of the Electrochemical Society*, 134(12) (1987), 2966.
18. D.D. Macdonald et al., Evaluation of alloy anodes for aluminum-air batteries: Corrosion studies, *Corrosion Science*, 44(9) (1988), 652.
19. C.D.S. Tuck, J.A. Hunter, and G.M. Scamans, The electrochemical behavior of Al–Ga alloys in alkaline and neutral electrolytes, *Journal of the Electrochemical Society*, 134(12) (1987), 2970.
20. W. Wilhelmsen et al., The electrochemical behaviour of Al in alloys in alkaline electrolytes, *Electrochimica Acta*, 36(1) (1991), 79.
21. P.W. Jeffrey, W. Halliop, and F.N. Smith, Aluminium anode alloy, EP209402A1 (1987).
22. E.J. Rudd and D.W. Gibbons, High energy density aluminum/oxygen cell, *Journal of Power Sources*, 47(3) (1994), 329.
23. M. Nestoridi et al., The study of aluminium anodes for high power density Al/air batteries with brine electrolytes, *Journal of Power Sources*, 178(1) (2008), 445.
24. V. Neburchilov et al., A review on air cathodes for zinc–air fuel cells, *Journal of Power Sources*, 195(5) (2010), 1271.
25. C.-C. Chang and T.-C. Wen, An investigation of thermally prepared electrodes for oxygen reduction in alkaline solution, *Materials Chemistry and Physics*, 47(2–3) (1997), 203.

26. E. Yeager, Electrocatalysts for O_2 reduction, *Electrochimica Acta*, 29(11) (1984), 1527.

27. L.Q. Hoa et al., Direct energy extraction from brown macroalgae–derived alginate by gold nanoparticles on functionalized carbon nanotubes, *ChemCatChem*, 6(1) (2014), 135.

28. J. Xu et al., Three-dimensional structural engineering for energy-storage devices: From microscope to macroscope, *ChemElectroChem*, 1(6) (2014), 975.

29. N.M. Marković et al., Oxygen reduction reaction on Pt and Pt bimetallic surfaces: A selective review, *Fuel Cells*, 1(2) (2001), 105.

30. M. Yuasa et al., Reverse micelle-based preparation of carbon-supported $La_{1-x}Sr_x$ $Mn_{1-y}Fe_yO_{3+\delta}$ for oxygen reduction electrode, *Journal of the Electrochemical Society*, 151(10) (2004), A1690.

31. Y. Shimizu et al., Bi-functional oxygen electrode using large surface area $La_{1-x}Ca_x$ CoO_3 for rechargeable metal-air battery, *Journal of the Electrochemical Society*, 137(11) (1990), 3430.

32. S. Müller et al., A study of carbon-catalyst interaction in bifunctional air electrodes for zinc-air batteries, *Journal of New Materials for Electrochemical Systems*, 2 (1999), 227.

33. M. Bursell, M. Pirjamali, and Y. Kiros, $La_{0.6}Ca_{0.4}CoO_3$, $La_{0.1}Ca_{0.9}MnO_3$ and $LaNiO_3$ as bifunctional oxygen electrodes, *Electrochimica Acta*, 47(10) (2002), 1651.

34. M. Hayashi et al., Enhanced electrocatalytic activity for oxygen reduction over carbon-supported $LaMnO_3$ prepared by reverse micelle method, *Electrochemical and Solid-State Letters*, 1(6) (1998), 268.

35. M. Hayashi et al., Reverse micelle assisted dispersion of lanthanum manganite on carbon support for oxygen reduction cathode, *Journal of the Electrochemical Society*, 151(1) (2004), A158.

36. J. Lukaszewicz et al., New approach towards preparation of efficient gas diffusion-type oxygen reduction electrode, *Journal of Materials Science*, 41(19) (2006), 6215.

37. L. Guangchuan et al., Influence of alkali metal cation (Li (I), Na (I), K (I)) on the construction of chiral and achiral heterometallic coordination polymers, *Journal of Rare Earths*, 25(Supplement 2) (2007), 264.

38. M. Yuasa et al., High-performance oxygen reduction catalyst using carbon-supported La-Mn-based perovskite-type oxide, *Electrochemical and Solid-State Letters*, 14(5) (2011), A67.

39. L. Wang et al., $CoMn_2O_4$ spinel nanoparticles grown on graphene as bifunctional catalyst for lithium-air batteries, *Journal of the Electrochemical Society*, 158(12) (2011), A1379.

40. T. Takeguchi et al., Layered perovskite oxide: A reversible air electrode for oxygen evolution/reduction in rechargeable metal-air batteries, *Journal of the American Chemical Society*, 135 (2013), 11125.

41. N. Li et al., Electrocatalytic activity of spinel-type oxides $LiMn_{2-x}M_xO_4$ with large specific surface areas for metal-air battery, *Journal of Power Sources*, 74(2) (1998), 255.

42. N. Li et al., Spinel-type oxides $LiMn_{2-x}M_xO_4$ [M=Co, Fe, (CoFe)] as electrocatalyst for oxygen evolution/reduction in alkaline solution, *Journal of Applied Electrochemistry*, 29(11) (1999), 1351.

43. T. Nissinen et al., Microwave synthesis of catalyst spinel $MnCo_2O_4$ for alkaline fuel cell, *Journal of Power Sources*, 106(1–2) (2002), 109.

44. Y. Kiros, Cathodes à diffusion d'air pour piles à combustible, WO 02/075827 A1 (2002).

45. J.S. Spendelow and A. Wieckowski, Electrocatalysis of oxygen reduction and small alcohol oxidation in alkaline media, *Physical Chemistry Chemical Physics*, 9(21) (2007), 2654.

46. B.B. Blizanac, P.N. Ross, and N.M. Markovic, Oxygen electroreduction on Ag(111): The pH effect, *Electrochimica Acta*, 52(6) (2007), 2264.

47. F. Bidault and A. Kucernak, A novel cathode for alkaline fuel cells based on a porous silver membrane, *Journal of Power Sources*, 195(9) (2010), 2549.

48. M. Yuasa et al., Bi-functional oxygen electrodes using $LaMnO_3/LaNiO_3$ for rechargeable metal-air batteries, *Journal of the Electrochemical Society*, 158(5) (2011), A605.

49. L. Mao et al., Mechanistic study of the reduction of oxygen in air electrode with manganese oxides as electrocatalysts, *Electrochimica Acta*, 48(8) (2003), 1015.

50. F. Cheng et al., MnO_2-based nanostructures as catalysts for electrochemical oxygen reduction in alkaline media, *Chemistry of Materials*, 22(3) (2010), 898.

51. J. Suntivich et al., Design principles for oxygen-reduction activity on perovskite oxide catalysts for fuel cells and metal–air batteries, *Natural Chemistry*, 3(7) (2011), 546.

52. F. Cheng and J. Chen, ChemInform abstract: Metal-air batteries: From oxygen reduction electrochemistry to cathode catalysts, *Chemical Society Reviews*, 41(6) (2012), 2172.

53. D. Higgins et al., Activated and nitrogen-doped exfoliated graphene as air electrodes for metal–air battery applications, *Journal of Materials Chemistry A*, 1(7) (2013), 2639.

54. S. Xu et al., A novel cathode catalyst for aluminum-air fuel cells: Activity and durability of polytetraphenylporphyrin iron (II) absorbed on carbon black, *International Journal of Hydrogen Energy*, 39(35) (2014), 20171.

55. D.B. Meadowcroft, Low-cost oxygen electrode material, *Nature*, 226(5248) (1970), 847.

56. R.S.M. Patnaik et al., Heat management in aluminium/air batteries: Sources of heat, *Journal of Power Sources*, 50(3) (1994), 331.

57. M.A. Rahman, X. Wang, and C. Wen, High energy density metal-air batteries: A review, *Journal of the Electrochemical Society*, 160(10) (2013), A1759.

58. M. Akmal, R. Othman, and M.H. Ani, Comparative electrochemical performance characteristics of aluminium-air cell employing seawater and NaCl electrolytes, *Advanced Materials Research*, 701 (2013), 314.

59. H. Arai and M. Hayashi, Secondary batteries—Metal-air systems I Overview (secondary and primary), in *Encyclopedia of Electrochemical Power Sources* (Eds. Chris K. Dyer, Patrick T. Moseley, Zempachi Ogumi, David A. J. Rand, Bruno Scrosati, Jürgen Garche), Elsevier, Amsterdam, the Netherlands (2009), p. 347.

60. D. Chu and R.F. Savinell, Experimental data on aluminum dissolution in KOH electrolytes, *Electrochimica Acta*, 36(10) (1991), 1631.

61. G.M. Wu, S.J. Lin, and C.C. Yang, Preparation and characterization of PVA/PAA membranes for solid polymer electrolytes, *Journal of Membrane Science*, 275(1–2) (2006), 127.

62. Z. Zhang et al., All-solid-state Al–air batteries with polymer alkaline gel electrolyte, *Journal of Power Sources*, 251 (2014), 470.

63. G.M. Wu, S.J. Lin, and C.C. Yang, Alkaline Zn-air and Al-air cells based on novel solid PVA/PAA polymer electrolyte membranes, *Journal of Membrane Science*, 280(1–2) (2006), 802.

64. T. Jiang et al., Electrodeposition of aluminium from ionic liquids: Part I—electrodeposition and surface morphology of aluminium from aluminium chloride (AlCl 3)–1-ethyl-3-methylimidazolium chloride ([EMIm] Cl) ionic liquids, *Surface and Coatings Technology*, 201(1–2) (2006), 1.

65. T. Hibino, K. Kobayashi, and M. Nagao, An all-solid-state rechargeable aluminum–air battery with a hydroxide ion-conducting Sb (V)-doped SnP$_2$O$_7$ electrolyte, *Journal of Materials Chemistry A*, 1(47) (2013), 14844.

66. R. Mori, A novel aluminium–air rechargeable battery with Al$_2$O$_3$ as the buffer to suppress byproduct accumulation directly onto an aluminium anode and air cathode, *RSC Advances*, 4(57) (2014), 30346.

67. L. Wang et al., A hybrid aluminum/hydrogen/air cell system, *International Journal of Hydrogen Energy*, 38(34) (2013), 14801.

68. A. Mukherjee and I.N. Basumallick, Metallized graphite as an improved cathode material for aluminium/air batteries, *Journal of Power Sources*, 45(2) (1993), 243.

69. C.-H. Wu et al., Inhibiting ability of chelating agent on aluminum corrosion in alkaline solution, *ECS Transactions*, 50(25) (2013), 23.

70. D.D. Macdonald and C. English, Development of anodes for aluminium/air batteries—solution phase inhibition of corrosion, *Journal of Applied Electrochemistry*, 20(3) (1990), 405.

71. X.Y. Wang et al., The effects of polyethylene glycol (PEG) as an electrolyte additive on the corrosion behavior and electrochemical performances of pure aluminum in an alkaline zincate solution, *Materials and Corrosion*, 61 (2010), 1.

72. A.M. Abdel-Gaber et al., Novel package for inhibition of aluminium corrosion in alkaline solutions, *Materials Chemistry and Physics*, 124 (2010), 773.

73. N.A.F. Al-Rawashdeh and A.K. Maayta, Cationic surfactant as corrosion inhibitor for aluminum in acidic and basic solutions, *Anti-Corrosion Methods and Materials*, 52 (2005), 160.

74. D.E. Fenton, J.M. Parker, and P.V. Wright, Complexes of alkali metal ions with poly(ethylene oxide), *Polymer*, 14(11) (1973), 589.

75. P.V. Wright, Electrical conductivity in ionic complexes of poly(ethylene oxide), *British Polymer Journal*, 7(5) (1975), 319.

76. M. Armand et al., Poly-ethers as solid electrolytes, in *Fast Ion Transport in Solids* (Eds. Vashishta, P., Mundy, J.N., and Shenoy, G.K.), North Holland, Amsterdan, the Netherlands (1979), pp. 131–134.

77. J.F. Fauvarque et al., Alkaline poly(ethylene oxide) solid polymer electrolytes. Application to nickel secondary batteries, *Electrochimica Acta*, 40(13–14) (1995), 2449.

78. S. Guinot et al., A new class of PEO-based SPEs: Structure, conductivity and application to alkaline secondary batteries, *Electrochimica Acta*, 43(10–11) (1998), 1163.

79. A.A. Mohamad, Electrochemical properties of aluminum anodes in gel electrolyte-based aluminum-air batteries, *Corrosion Science*, 50(12) (2008), 3475.

80. R. Mori, A new structured aluminium-air secondary battery with a ceramic aluminium ion conductor, *RSC Advances*, 3(29) (2013), 11547.

81. R. Mori, A novel aluminium-air secondary battery with long-term stability, *RSC Advances*, 4(4) (2014), 1982.

82. N.S. Hudak, Chloroaluminate-doped conducting polymers as positive electrodes in rechargeable aluminum batteries, *The Journal of Physical Chemistry C*, 118(10) (2014), 5203.

83. M. Parans Paranthaman et al., Chapter 13—Aluminum-ion batteries for medium- and large-scale energy storage, in *Advances in Batteries for Medium and Large-Scale Energy Storage* (Eds. C. Menictas, M. Skyllas-Kazacos, T. M. Lim), Woodhead Publishing, U.K. (2015), p. 463.

84. F. Shu et al., Research progress in Al alloy anodes for alkaline Al-air battery, *Battery Bimonthly*, 35(2) (2005), 158.

85. P. Quaino et al., Why is gold such a good catalyst for oxygen reduction in alkaline media? *Angewandte Chemie International Edition*, 51(52) (2012), 12997.

86. W. Wan, Y. Tang, and Z. Lu, Effect of additives in the alkaline electrolyte on anode behavior of aluminum, *Battery Bimonthly*, 38(1) (2008), 40.

87. C. Li et al., Metallic aluminum nanorods: Synthesis via vapor-deposition and applications in Al/air batteries, *Chemistry of Materials*, 19(24) (2007), 5812.

88. D. Gelman, B. Shvartsev, and Y. Ein-Eli, Aluminum–air battery based on an ionic liquid electrolyte, *Journal of Materials Chemistry A*, 2(47) (2014), 20237.

89. L. Fan and H. Lu, The effect of grain size on aluminum anodes for Al-air batteries in alkaline electrolytes, *Journal of Power Sources*, 284 (2015), 409.

4 Magnesium–Air Battery

Bruce W. Downing

CONTENTS

4.1 INTRODUCTION

The rapid growth of portable electronics has led to greatly expanded opportunities for small, lightweight consumer-friendly portable power systems.

Metal–air batteries [1–11] are one of the more promising, but less well-known, alternatives to conventional and future power sources as primary cells. Metal–air systems, such as the magnesium–air, are typically very high in energy density but low in power, have an open cell structure, and use oxygen from the air. Great strides have been made and are continuing to be made in the metal–air technology. They have the potential to replace conventional batteries (e.g., zinc alkaline) and high-cost hydrogen-based fuel cells because of their high energy density, relatively flat discharge voltage potential, long shelf life, and relatively low manufacturing cost.

Metal–air batteries are an attractive power source for stand-alone power supplies (e.g., for standby or emergency power and portable power). They feature the electrochemical coupling of a metal anode to an air diffusion cathode through a suitable electrolyte. This combination produces a cell with an inexhaustible cathode reactant from the oxygen in atmosphere air. Metal–air batteries also have the versatility of an aqueous or nonaqueous electrolyte.

Metal–air battery has also been coined a semi-fuel cell, which is a hybrid between a fuel cell and a battery; however, this term has not been used in the industry and is notably absent in any codes and standards publications. Though the term "fuel cell" has been used for metal–air technology, more so for marketing purposes when fuel cell was the public perception term, the metal–air technology is a closed system as opposed to a fuel cell, which is an open system. The continuous (open) energy carrier in a fuel cell is hydrogen while the metal anode energy carrier in a metal–air is stationary.

The simple cell structures of metal–air batteries can be adapted to almost any size for use in many applications and that the cell can be mechanically recharged in seconds and the cathode reaction uses oxygen, a readily available cost-free gas from the atmosphere.

The main advantages associated with metal–air batteries are as follows:

- High energy density (on the basis of metal and electrolyte)
- Flat discharge voltage
- Long shelf life (based on dry storage)
- Relatively low cost (on the basis of metal and air diffusion cathode used)

The power capacity of metal–air batteries is a function of several variables:

- Air diffusion cathode area
- Catalyst type used in the air diffusion cathode (i.e., Mn vs. Co)
- Air availability and oxygen content
- Humidity
- Anode type and anode alloy material (i.e., Al, Mg, Li, and Zn)
- Anode mass, morphology, and corrosion mechanisms
- Electrolyte type and concentration

- Additives
- Operational temperature
- CO_2 content of air
- Manufacturing quality and components

The magnesium–air battery technology does lend itself as a power source to portable electronic devices, and comparisons with other metal–air types are shown in Tables 4.1 and 4.2

The following treatise is based on closed magnesium–air systems and is therefore directed toward the saltwater-based aqueous electrolyte battery as there are more commercially available designs and products than for the nonaqueous battery. It is based largely on laboratory and prototype test work conducted by the present writer [12] and colleagues at MagPower Systems, Inc., White Rock, BC, Canada.

TABLE 4.1
Magnesium, Aluminum, and Zinc Metal–Air Types

	Magnesium	Aluminum	Zinc
SG	1.74	2.70	7.13
SHE (V)	−2.363	−1.662	−0.763
Energy # electrons	2	3	2
Open circuit voltage	1.7	1.2	1.3
Anode composition (wt%)	>90%	99.999%	99.99%
Current capacity (Ah/kg)	2200	2500 (alloy)	740
Electrolyte	Salt water	KOH	KOH
pH	6–8	13–14	13–14
Source(s)	Magnesite limestone, seawater	Bauxite	Zinc sulfides

TABLE 4.2
Properties of Metals Used in Metal–Air Batteries

Metal Anode	Electrochemical Equivalent of Metal (Ah/g)	Theoretical Cell Voltage (V)	Theoretical-Specific Energy (of Metal) (kWh/kg)
Li	3.86	3.4	13
Ca	1.34	3.4	4.6
Mg	2.2	3.1	6.8
Al	2.98	2.7	8.1
Zn	0.82	1.6	1.3
Fe	0.96	1.3	1.2

Source: Energy Storage, March/April, Refocus, Elsevier Advanced Technology Article, 2002.

4.1.1 HISTORY

The magnesium–air battery technology has been known since the 1960s [13–17], but no major advancement has been made toward commercialization due to technological barriers such as hydrogen generation, precipitate control, anode material, air diffusion cathode material, electrolyte composition, and expensive manufacturing processes.

Magnesium–air batteries have been discussed in recent publications, indicating the renewed interest in this technology [18–21].

The early adaptation of the technology was used in torpedoes during World War II, and sonobuoys, both of which used a seawater electrolyte. Applications are discussed in Section 4.8.

4.1.2 BATTERY TYPES

Magnesium–air batteries have the versatility of an aqueous or nonaqueous electrolyte. The most common type is the aqueous electrolyte; however, great strides have been made within the last decade to create nonaqueous (organic) electrolytes that consist of a gel polymer electrolyte membrane serving as both the separator and ion-transporting medium [22,23]. This also allows the battery to become rechargeable, albeit from an external source of power. Despite its advantages, the application of magnesium rechargeable batteries has had several problems regarding magnesium diffusion, passivating surface film, lack of great electrolytes [24], and low specific energy of the battery [19,25]. A review of the history, breakthroughs, and challenges regarding the magnesium rechargeable battery is presented by Yoo et al. [26]. This battery technology is highly promising for energy storage and conversion, and if manufacturing costs can be in the range of other comparable batteries, then the consumer has a wider range of battery products to choose from for specific high-energy devices such as electric cars.

This chapter on magnesium–air is directed toward the saltwater-based aqueous electrolyte battery as there are more commercially available products than the non-aqueous magnesium battery type.

4.1.2.1 Magnesium–Air Battery Technology

The magnesium–air battery system utilizes common materials such as magnesium, saline water, and air to provide a high-performance primary battery. A schematic of a typical magnesium–air cell (MAC) is shown in Figure 4.1. These components (i.e., magnesium, water, and air) are relatively inert, which provides a high degree of safety to the magnesium–air battery system. The anode of the magnesium–air battery system is a common magnesium metal (mixed with small amounts of other additives) that undergoes oxidation to produce Mg^{2+} and two electrons. Magnesium metal and alloys are extremely stable in air, unlike lithium battery anodes, and the stability and safety of this material has prompted its potential use in the airline industry. At the cathode of the battery system, oxygen taken from air combines with water to undergo reduction to produce hydroxyl ions. A saline electrolyte connects the anode and cathode to complete the electrochemical circuit. Removal of the

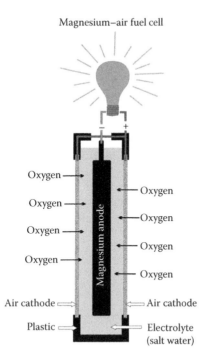

FIGURE 4.1 Magnesium–air cell (aqueous).

electrolyte suspends the chemical processes involved in converting chemical energy to electrical energy, and the anode and cathode materials become dormant. Because of the stability of the anode and cathode materials, the battery cell stack can remain unchanged for a significant period of time once the electrolyte has been removed (or before the electrolyte has been added). When the electrolyte is added back, the battery again begins to produce electricity.

The magnesium–air battery is rechargeable in that the anode (fuel) needs to be replaced mechanically when it is completely corroded. The electrolyte also needs to be either replenished and/or replaced as it is consumed and can be open to evaporation.

The performance of the magnesium–air battery system depends on the interaction of eight major components:

1. Cell assembly (configuration and mechanical + electrical integrity of the unit)
2. Electrolyte composition
3. Hydrogen suppressant additive to electrolyte (compound and concentration)
4. Air diffusion cathode structure and composition (catalyst type and loading)
5. Cell gap (initial + magnesium thickness)
6. Anode composition and morphology
7. Electrolyte pumping and heat exchange
8. Precipitate control

4.1.3 ELECTROCHEMICAL PROCESSES

In the magnesium–air technology, electricity is generated through the combination of magnesium, oxygen, and a saltwater (i.e., NaCl) electrolyte via the following reactions:

Electrons are formed at the anode:

$$\text{Anode } Mg^{2+} + 2e^- \leftarrow Mg \tag{4.1}$$

$$Mg + 2H_2O \; Mg(OH)_2 + H_2 + 353 \text{ kJ} \tag{4.2}$$

Equation 4.2 is also referred to as a "parasitic" reaction in that hydrogen is formed with release of heat energy (discussed in Section 4.9.1).

OH⁻ ions are produced at the cathode and electrons are consumed. An increase in pH will occur due to OH⁻ generation:

$$\text{Cathode } \frac{1}{2}O_2 + H_2O + 2e^- \; 2(OH)^- \tag{4.3}$$

$$Mg^{2+} + 2(OH)^- \; Mg(OH)_2 \tag{4.4}$$

The net cell electrochemical reaction is

$$\text{Magnesium} + \text{Water} + \text{Oxygen reaction by products} + \text{DC power} + \text{Heat}$$

The practical open cell voltage of an MAC is 1.7 V, which is based on actual measurement. The operating open cell voltage is the product of full cell reaction potential minus the electrode overpotentials and the ohmic voltage (IR) drop in electrolytes and separators.

The numerical value of the full theoretical reaction potential is 2.356 V (at 25°C). The formula is as follows:

Cathode reaction

$$O_2 + 2H_2O + 4e^- \; 4OH^- \; E^\circ = 0.40 \; V_{SHE} \tag{4.5}$$

Anode reaction

$$2Mg^{2+} + 4e^- \; 2Mg \; E^\circ = -2.756 \; V_{SHE} \tag{4.6}$$

(4.5)–(4.6)

$$O_2 + 2H_2O + 2Mg \; 2Mg(OH) \; E^\circ = 2.356 \; V_{SHE} \tag{4.7}$$

Passivation

The magnesium anode can become passivated by $MgO/Mg(OH)_2$.

The reaction is slow in pure water but is accelerated with high electrical conductive additives such as sodium chloride (NaCl). This reaction is exothermic,

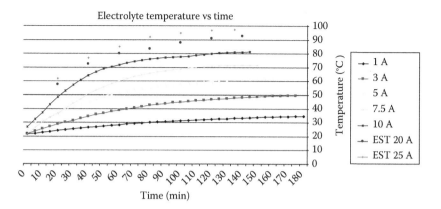

FIGURE 4.2 Effect of current and temperature using 12 cells (ambient temperature).

the temperature depending upon the load of the power device (i.e., the greater the amperage, the greater the amount of heat generated; Figure 4.2). The magnesium utilization efficiency (MUE) (discussed in Section 4.9.1) of the cell will in part depend upon the internal cell temperature with 70°C being the most efficient running temperature.

The electrons produced at the anode pass around the external circuit connected to the fuel cell, and both reactions (anode and cathode) carry on until the magnesium is entirely consumed. When the MAC is under load, it is normal for the water to be consumed and evaporated, thus decreasing the volume of the electrolyte solution. As the electrolyte volume decreases, the reaction product(s) mass increases, which can lead to electrolyte supersaturation and eventual precipitation of the reaction product(s) and thus clogging of the cell.

Magnesium exhibits an electrochemical corrosion phenomenon known as the negative difference effect (NDE). This effect has been the subject of research dating from the 1950s with varying authors [27,28] attempting to explain the NDE by means of electrochemical reaction mechanisms. Atrens [29] discusses the two mechanistic approaches to the magnesium NDE: (1) the unipositive Mg^+ ion NDE mechanism and (2) the magnesium self-corrosion mechanism. Liu [30] discusses the NDE behavior based on Tafel kinetics proposing that both Mg^+ and Mg^{2+} result from magnesium dissolution and Mg^{2+} increases faster than Mg^+ with the increase of the overpotential. From experimental work regarding the MAC (12), the decrease of magnesium efficiency with increased power output is probably a consequence of the anomalous NDE on magnesium. The NDE means that the rate of parasitic generation of hydrogen rises when the anode undergoes higher positive polarization as the current density is increased.

4.1.4 MAGNESIUM–AIR BATTERY COMPONENTS

A magnesium–air battery is made up of one or more MACs combined in series depending upon the required power output. The battery can be constructed so as the electrolyte can be in a (1) stagnant mode, (2) pumped mode using an electronic pump, or (3) pumped mode using a mechanical pressure differential system.

The magnesium–air battery employs a metal (anode) such as magnesium or a magnesium-based alloy as the fuel immersed in an electrolyte coupled with a gas diffusion cathode (GDE) to generate direct current. The aqueous or nonaqueous electrolyte is a conducting medium in which the migration of ions creates a flow of electric current.

The GDE consists of (a) a current collector; (b) reactive carbon layers, one on either side of the current collector with or without a catalyst; and (c) a microporous hydrophobic film on the air side to allow oxygen-laden air to pass through to the electrolyte. The GDE can range from a one- to four-layer structure and serves to provide an interface where oxygen from air is dissolved in an electrolyte solution and catalytically reduced on the active components of the GDE, which is carbon with or without a catalyst to enhance the rate of oxygen reduction (oxidation reduction reaction). The cathode is the positive electrode by which the flow of electrons enters. The structure and chemistry of the air cathode is very important, and there are many designs, some of which are commercial and others that are the priority of the battery manufacturer.

The fuel (anode) is magnesium that provides the electrons that can be deemed as a sacrificial anode and is replaceable. The metal (anode) is the negative electrode by which the current leaves. During the exothermic reaction, the metal anode undergoes corrosion, and the electrolyte is consumed and broken down into separate components that forms the reaction products:

$$\text{Metal (fuel)} + \text{Electrolyte} + \text{Oxygen} = \text{Direct current} + \text{Reaction product(s)} + \text{Heat}$$

Reaction products will depend upon both the chemistry of the magnesium metal and the electrolyte.

4.1.5 MAGNESIUM–AIR BATTERY PERFORMANCE

Assembly, testing, and diagnosis of the MAC has been conducted in several laboratories using different types of equipment. Figures 4.3 through 4.8 show typical power performances of the MAC; however, it should be noted that the cells used in the test work do not represent any new and improved MAC versions that the author may have access to. This test work was conducted using a Thixotech fabricated AM60B magnesium anode.

The steep drop in cell potential at low current densities is due in large part to the polarization of the air electrode (cathode); the linear drop at higher current densities is mainly the internal resistance drop of the air electrode and electrolyte. After the initial polarization, the magnesium electrode (anode) shows essentially a constant potential over the current range shown.

The common cause of an unreasonable load is a short circuit of the battery or fuel cell. For example, a lead-acid battery will be damaged, lose its rechargeable capacity, and will experience a shortened lifetime if a short circuit occurs; this is due to generated heat and shedding of paste. Unlike the lead-acid battery, the MAC will

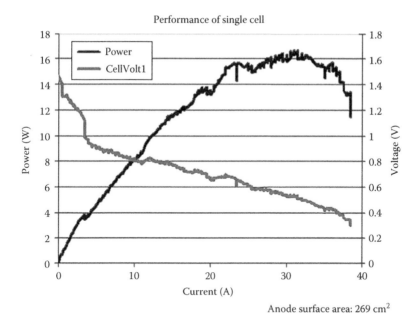

FIGURE 4.3   Power graph, 1 cell. (From NRC Fuel Cell Institute, May 2004.)

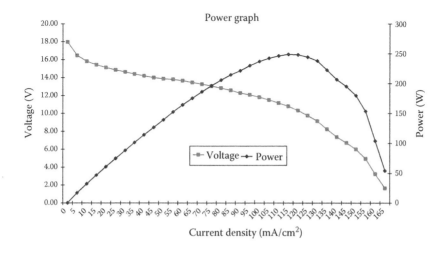

FIGURE 4.4   Current density data, 12 V MAFC (12 cell stack). (From Vizon Scitec, Inc., October 2004.)

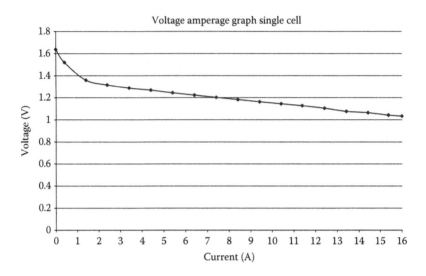

FIGURE 4.5   Voltage versus current (single cell). (From MagPower, January 2005.)

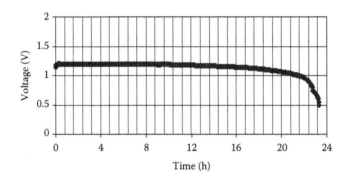

FIGURE 4.6   Discharge curve—1 cell at 5 A. Energy = 108 Wh, Temp = 55°C, vol = 0.5 L, wt anode = 140 g. (From Vizon Scitec, Inc., July 2003.)

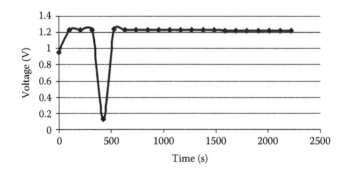

FIGURE 4.7   Short circuit—1 cell (5 A discharge) short = 212 s. (From Vizon Scitec, Inc., July 2003.)

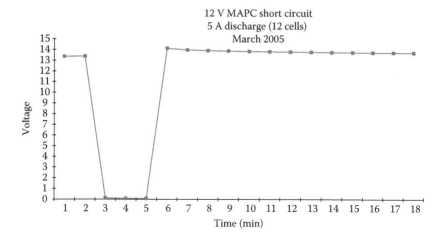

FIGURE 4.8 Short circuit—12 cells (5 A discharge). (From MagPower, March 2005.)

only lose some capacity due to the waste consumption of the magnesium anode during the short circuit. No damage of the MAC components, or loss of future capacity, will be affected. The following graph shows the response of a single MAFC cell under short circuit conditions. The MAC recovers to its operating voltage right after the short is dissolved.

During one of the test programs, the load was inadvertently reversed (i.e., the cell was in fact "charged" instead of "discharged"). Figure 4.9 shows the current and voltage reaction during running power curve sequence to 10 A with 0.25 A/30 s step.

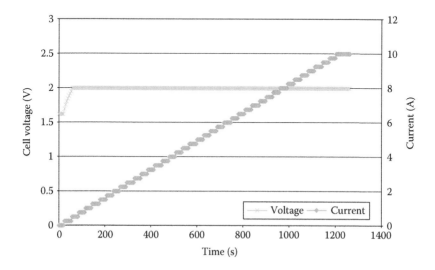

FIGURE 4.9 Current and voltage reaction during running power curve sequence to 10 A with 0.25 A/30 s step.

When the current reached 0.5 A, the voltage increased to 2 V. The voltage should rise with increased current; however, the test stand was limited to measure voltage up to 2 V. It lasted 20 min to reach 10 A.

After the aforementioned sequence, the polarity connected to load bank was corrected. The cell was discharged at 5 A for 112 s, and the cell potential was around 1.1 V (as shown in Figure 4.10). Subsequently, a power curve up to 10 A with 0.25 A/30 s step was obtained and is shown in Figure 4.11. Based on the aforementioned results, the charged cell did not exhibit obvious degradation compared with the normal cells. There was also no safety issues.

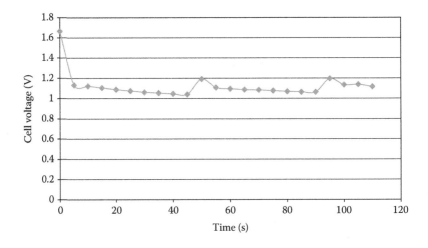

FIGURE 4.10   Cell voltage change with time when the cell was discharged at 5 A.

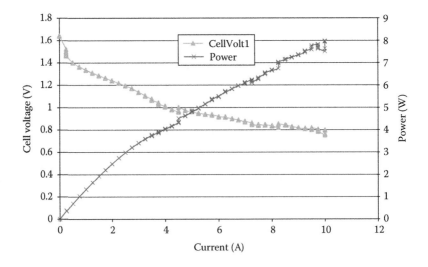

FIGURE 4.11   Cell voltage and power changes with discharging current.

4.2 ANODE

The aqueous magnesium–air battery employs a magnesium or magnesium alloy anode as the fuel.

4.2.1 MAGNESIUM

Magnesium is one of the lightest metals that together with magnesium alloys have properties that can lead to specific applications. Magnesium (Mg—atomic number 12) was first isolated by the British chemist Sir Humphrey Davey in 1808. The name originates from the Greek word for a district in Thessaly called Magnesia. Magnesium is the eighth most abundant element in the Earth's crust, and although it is not found in its elemental form, it is found in 663 minerals as a component of the mineral (i.e., brucite—$Mg(OH)_2$). Magnesium is also the third most plentiful element dissolved in seawater, with average concentration of 1326 mg/L, but can be quite variable in freshwater due to groundwater interactions with different surficial media. Seawater will vary in trace elements, temperature, salinity, and conductivity depending upon the vertical gradient (depth) and spatial area in the world. Magnesium exceeds calcium in abundance in seawater but is far less common than calcium in freshwater. Where magnesium content is quite high in freshwater, it occurs naturally in hard water, which is essentially water that has come in contact with geological material such as dolomite that has a high content of magnesium. In this situation, water softeners are generally purchased that remove calcium and magnesium via ion-exchange resin and back flushed into water sewage lines using a salt brine solution. There are no particular standards or guidelines for magnesium in drinking water as compared to Al and Zn and are generally site specific.

Bioavailable magnesium is an important element for plant and animal life, and the adult human daily requirement of magnesium is about 0.3 g per day. Magnesium is an essential element that is involved in energy metabolism, protein and nucleic acid synthesis, formation of bones, and activity of neurotransmitters. When dissolved in water, magnesium carries two positive charges (Mg^{2+}), and because of this relatively high charge has major importance as an electrolyte. Magnesium causes no unwanted side effects for most people and toxicity is rare. Numerous foods contain magnesium in varying amounts ranging from 13 mg (milk) to 490 (wheat bran). Information for this section is taken from articles published on various websites (see references—health). Magnesium metal is noncarcinogenic and does not have well-characterized toxicity. Magnesium compounds are generally noncarcinogenic and nontoxic.

Magnesium occurs fifth in the electrochemical series, meaning that any metal higher up in the series will have a greater potential difference (relative to the standard hydrogen electrode [SHE]) than those below (i.e., Mg is more electrochemically reactive in saltwater than either Al or Zn and has a greater electropotential than either of these metals) (Table 4.3). This galvanic corrosion chart says that the "anodic" or "less noble" metals at the negative end of the series—at the right of this diagram, such as magnesium, zinc, and aluminum—are more likely to be attacked than those at the "cathodic" or "noble" end of the series such as gold and graphite.

TABLE 4.3
Electrochemical (Galvanic Corrosion) Chart

Metal	Symbol	
Potassium	K	Most electropositive element, anodic, less noble
Sodium	Na	
Calcium	Ca	
Magnesium	**Mg**	
Aluminum	Al	
Zinc	Zn	
Iron	Fe	
Lead	Pb	
Hydrogen	*H*	
Copper	Cu	
Mercury	Hg	
Silver	Ag	
Gold	Au	
Graphite	C	Least electropositive metal, cathodic, noble

Note: Bold shows the placement of magnesium in the table and italic shows the placement of hydrogen in the table.

4.2.1.1 Sources

Magnesium metal can be recovered from seawater and lake brines, as well as from minerals such as dolomite (calcium–magnesium carbonate), serpentine (hydrous magnesium silicate), olivine (magnesium silicate), magnesite (magnesium carbonate), and brucite (magnesium hydroxide). Recycling has now become an additional source of magnesium metal. Major magnesium-producing countries include People's Republic of China, Australia, Canada, and Russia.

4.2.1.2 Production

Magnesium can be made commercially by several processes and can be recovered from seawater (Dow Chemical). There are two major process routes utilized for production of magnesium metal, namely, electrolytic route and thermal route.

The technologies for the production of magnesium by electrolysis (electrolytic route) are characterized by three essential steps:

- Production of magnesium-rich brine (usually $MgCl_2$ or $MgCl_2/KCl$)
- Brine drying or crystallization followed by dehydration
- Electrolytic reduction of $MgCl_2$ or $MgCl_2$*-KCl for production of Mg-metal and chlorine gas

Nonelectrolytic methods (thermal route) of magnesium production are all based upon thermochemical reduction of magnesium oxide, which is obtained by processing any one of a number of magnesium resources, for example, magnesite, dolomite, and olivine, although dolomite is utilized most commonly. Newer processes such as Noranda's Magnola [31] is essentially a leaching process of serpentinite material but with high operational costs. The most important factor is the choice of a process that maximizes energy efficiency, in addition to locating a plant where energy sources are not only readily available, but also at favorable rates.

4.2.1.3 Uses

Magnesium is a metal that has both major industrial and health uses and is becoming more important as a result of new technology and markets, especially in the automotive industry. It is the lightest structural metal known. Magnesium is a silvery white metal that is strongly reactive with oxygen at high temperatures (above 645°C) in dry air; however, it will form a stable water-insoluble magnesium oxide/hydroxide film on the metal's surface at room temperature.

- Since it is 30% lighter (density of 1.738 g/cm^3 and atomic weight of 24.3050 g/mol) than aluminum, it is used in products and alloys such as
  - Aircraft and missile construction
  - Automotive industry
  - Artificial limbs
- A reducing agent for the production of uranium and other metals from their salts
- An anode in hot water tanks
- In flares and pyrotechnics, including incendiary bombs
- In computers for radiofrequency shielding
- In numerous compounds
  - Used to make organomagnesium compounds (Grignard reagents), useful in organic synthesis
  - Magnesium oxide
    - Since it is refractory, it is used as bricks and liners in furnaces.
  - Magnesium hydroxide
    - Acid neutralizer (i.e., acid rock drainage)
    - Fire retardant
    - Cosmetic preparations as an absorbent and pH adjuster
    - Milk of magnesia, Maalox (Mg Al hydroxide)
  - Magnesium sulfate
    - Epsom salts
  - Magnesium citrate
    - Medicine
    - Effervescent beverages
  - Magnesium chloride
  - Vitamins/minerals (a major element in mineral tablets)

4.2.1.4 Associations

The International Magnesium Association (IMA) is a worldwide organization, headquartered in McLean, Virginia, embracing the interests of those concerned with the metal magnesium. Founded in 1943, the IMA strives to promote the magnesium industry by collecting and disseminating information, by encouraging research, and by publishing innovative uses of the metal. The magnesium website can be accessed via www.intlmag.org, which discusses in detail the health aspects of magnesium. Another organization dedicated to magnesium can be found at www.magnesium.com.

4.2.2 Magnesium Anode

Understanding the electrochemical corrosion behavior and mechanisms of magnesium is important to the commercialization of the magnesium–air battery. An important and limiting factor to increase power density in the magnesium–air power source is the magnesium anode itself (albeit in conjunction with the electrolyte and air cathode). The anode is the fuel (or fuel carrier) and is consumed and must be replaced. The mass of the anode will influence both the run time and amperage of the MAC. The metal quality of the anode is paramount to the MUE. The metal quality [32] may be defined as (1) chemical composition, (2) inclusions and porosity inside the anode, (3) the surface appearance, and (4) consistency. Magnesium anodes are generally specified with respect to strict chemical compositions because of harmful impurities such as iron, copper, cobalt, and nickel, which can impair the corrosion performance of the anode and lead to nonuniform dissolution. These impurities can result in generating local galvanic cells on the anode, which has an effect of reducing anode efficiency due to their low solid-solubility limits and their impact as active cathodic sites for the reduction of water at the sacrifice of elemental magnesium [33]. Aluminum, manganese [34], and zinc can be added to promote uniform corrosion and reduce the rapid activation of magnesium. Test work to date has indicated that a few commercially available magnesium and magnesium alloy anodes can be used in the MAC, but that they have some limitations. They were essentially developed for other commercial applications such as automobiles, cathodic protection for boats, and seawater batteries. The present anodic structure is sheeted and is essentially two-dimensional. The current anode compositions that are commercially available may not be the best for this power source. Thus, it is possible to further improve the performance of the magnesium–air power source anode in order to achieve useful power for commercial applications.

Understanding corrosion behavior and mechanisms of magnesium [35–38] provides a basis for the design of new alloys and manufacturing techniques. Microgalvanic corrosion reaction between magnesium, alloy elements/impurities, and grain size is important and forms the understanding of passivation and dissolution of magnesium [30,39]. Another factor to consider in the magnesium alloys is the presence of various phases that can have an impact upon corrosion behavior [37].

Several types of anodes have been tested together with different manufacturing methods in prototype cells such as Mg, MgMn (1 wt% Mn), AZ31, AT61, AM50, and AM60B. The current Mg alloys are designed for their mechanical or physical properties, and they may not be the best alloys for use as an anode in the MAFC. For example, Al and Mn in the Mg alloys are known to reduce the long-term dissolution rate of the metal in various solutions. The current density of Mg anode can be increased by alloying Mg with elements that increase the dissolution rate. Based on current literature, it may be feasible to design the anode chemistry so that micro-galvanic cells are formed as current is drawn out of the fuel cell battery. The anode chemistry should also be assessed as to any impacts it will have on the environment (i.e., eco-material assessment), which is primarily trace element composition. From test work and literature review, the chemistry, especially trace element content, is important in the development of the anode for the MAC. It is imperative that the correct trace element chemistry be initiated and that high purity with very low impurity concentrations is attained. Trace elements and concentrations are ultimately influenced by the raw material at source.

From test work and literature review, the metallurgical factors such as impurities, alloy composition, and grain size influence the manufacturing process, which is extremely important in the development of the anode for the MAC (Figure 4.12). This is evident in the grain alignment, porosity, and grain compaction. This in turn impacts the surface quality of the anode (i.e., caking of precipitate on anode) and precipitate quality and color (Figure 4.13). It is imperative that the correct manufacturing process be initiated as it will influence the structural integrity of the anode material. The magnesium anode commercial manufacturing methods/processes used to date are die casting, thixomolding, extrusion, and cold-pressing metal powder. It was found that both thixomold and die cast have a greater amount of porosity than extruded material and that during casting, flow lines and uneven surfaces in the anode have a detrimental effect on corrosion behavior. The exposed surface area of the anode is important as it will impact the reaction integrity. Where large currents coupled with a high rate of magnesium corrosion are required, then perforated or foam anodes will have more reaction surfaces but at the same time be consumed

FIGURE 4.12 Die cast AZ31, thixomold AM60B, die cast AZ31.

FIGURE 4.13   Extruded anode material, Mg, MgMn, and AZ31.

faster than a solid anode. Magnesium anode micro-/nanostructures [18] have been examined that they produce high volume of surface active sites, resulting in better performance in comparison to solid anode structures.

Another type of anodic material is cold-pressed magnesium powder. This type of anode that results in a porous microstructure with increased surface area greatly increases the power density but results in increased dissolution rate of magnesium. Experimental work and fabrication development using different magnesium powders (Figures 4.14 and 4.15) [40] were developed for a cost-effective and simple fabrication process to manufacture the powder anode for magnesium–air battery with improved performance. This process makes it convenient to incorporate different alloying additions, as well as additives (Figure 4.16). The anode composition was designed so that microgalvanic cells were formed as current is drawn out of the fuel cell, resulting in increased dissolution rate of Mg. Furthermore, quantitative kinetics analysis of the obtained I–V curves and electrochemical impedance spectroscopy were carried out to clarify the mechanism.

A limiting factor is the retention of the salt water electrolyte within the pore spaces, which leads to constant discharge even when the electrolyte is removed.

Another type of magnesium anode that may have some commercial promise is magnesium powder coated on a reticulated carbon electrode.

The preferred anode in test work to date is extruded AZ31. The apparent reason that magnesium hydroxide does not cake on the AZ31 versus other anodes is that this material causes a decrease in the magnesium reaction activation as compared to using pure magnesium material, which has a high magnesium reaction activation (thus the reason for cleaner anodes and cathodes), which may be due to the presence of zinc. No dendrites were noted during any of the experimental test work.

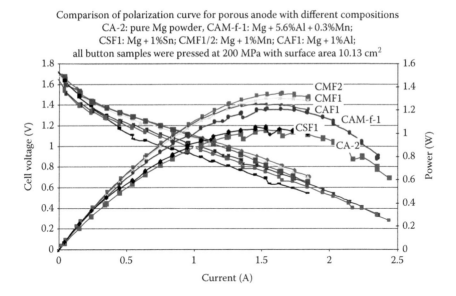

FIGURE 4.14 Comparison of anode powders pressed at 200 MPa, surface area 10.13 cm² (CA2, 100% Mg; CMF1 and 2, 99% Mg, 1% Mn; CAF1, 99% Mg, 1% Al; CAMf-1, 94% Mg, 5.6% Al, 0.3% Mn; CSF1, 99% Mg, 1% Sn).

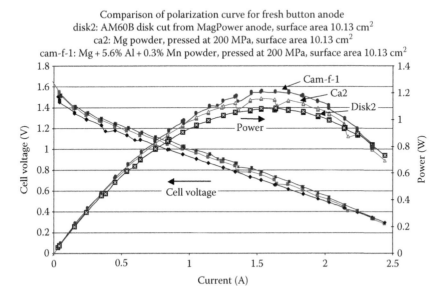

FIGURE 4.15 Comparison of anode powders pressed at 200 MPa, surface area 10.13 cm² (CA2, 100% Mg; CAMf-1, 94% Mg, 5.6% Al, 0.3% Mn; disk2: AM60B, 94% Mg, 6% Al; CSF1, 99% Mg, 1% Sn).

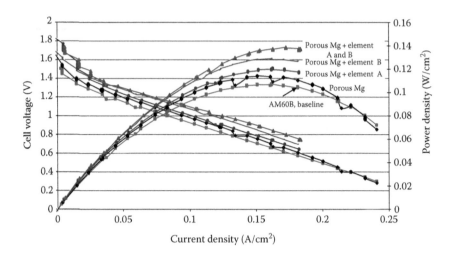

FIGURE 4.16 Polarization curves of porous anodes with different alloying additions versus solid AM60B as baseline. Elements A and B are intellectual property of MagPower Systems, Inc.

4.2.2.1 Reaction Products

Reaction products (precipitates) will occur from the natural reaction of the anode immersed in the electrolyte. These reaction products will vary depending upon the electrolyte and anode compositions. The dominant reaction products as identified from x-ray diffraction method are brucite (magnesium hydroxide—$Mg(OH)_2$) and iowaite ($Mg_4Fe(OH)_8OCl-4H_2O$). Other minerals identified in minor concentrations include aragonite, burkeite, chlorartinite, thenardite, and a magnesium chloride hydroxide hydrate. Other chemical species such as magnesium hydride and magnesium chloride will occur as deduced from overall chemical reactions but have not been identified via x-ray diffraction probably due to very low concentrations. Halite (NaCl) will precipitate when the saturation point of the reaction products in the aqueous electrolyte is reached. If carbon dioxide were present, then magnesium hydroxide could combine with carbon dioxide to form magnesium carbonate. An excellent method of determining the various reaction products is from the use of Pourbaix diagrams and x-ray diffraction determinations of the precipitates.

Ideally, if hydrochloric acid was added to the magnesium hydroxide saturated electrolyte, then magnesium chloride and water would be formed. This reaction could then extend the life of the aqueous electrolyte with the addition of water. Because H^+ forms an acid and OH^- forms a base, this is an acid–base reaction, and salt and water will be produced as products.

$$\text{Balanced equation: } 2HCl + Mg(OH)_2 \rightarrow MgCl_2 + 2H_2O$$

$$(4.8)$$

$$\text{Ionic equation: } 2H+(aq) + 2OH-(aq) \rightarrow 2H_2O$$

Mg and Cl do not appear in the ionic equation since they are spectator ions.

Investigation of electrowinning magnesium from the magnesium hydroxide saturated electrolyte is not cost effective and thus not feasible.

4.3 CATHODE

The air diffusion cathode is a sheet-like member having opposite faces exposed to two different environments, an atmosphere and an aqueous solution or an atmosphere and a solid, respectively. The air diffusion cathode must form a three-phase (gas–solid–liquid) interface where gas, catalyst/carbon, and electrolyte are in contact to facilitate the reaction of gaseous oxygen. The atmospheric side needs to be permeable to air but substantially hydrophobic in order to avoid electrolyte leakage through the air diffusion cathode to the atmosphere boundary, and reactant flooding. The current collector embedded in the air diffusion cathode is necessary for current flow and a structural support for the air diffusion cathode. During operation, oxygen passes through the air-permeable and hydrophobic air diffusion cathode by difference in pressure between the outside and inside of the cell. The catalyst within the carbon cathode facilitates the reduction of oxygen to anion via electrochemical reaction with electrons generated from the oxidation of the magnesium anode. The hydroxyl ions migrate from the air cathode to the magnesium anode to complete the cell reaction.

The discharge reaction mechanism of a metal–air battery is expected as follows if the cathode O_2 reduction is a four-electron process:

$$\text{Anode Metal} \rightarrow \text{Metal}^{\frac{2+}{3+}} + \frac{2e^-}{3e^-},$$

$$\text{Cathode } \frac{1}{2}O_2 + H_2O + 2e^- \rightarrow 2OH^-,$$

When the battery is in operation, air is drawn into and through the air diffusion cathodes. The oxygen will be used in the oxidation-reduction reaction between the air cathode and magnesium anode. Both oxygen and nitrogen gases pass through the cathode gas diffusion layer together and react with the catalyst layer, oxygen will be reduced to H_2O_2, and nitrogen gas will diffuse into the electrolyte and thus form bubbles in the electrolyte and be released into the air. The electrolyte is composed of water that contains air by nature. Air is composed of approximately 21 volume% oxygen and 78 volume% nitrogen with the other 1% made up of other gasses such as CO_2. When the water is agitated by pumping mechanism, then any air bubbles will disperse from the electrolyte. Hydrogen is generated as a secondary reaction on the anode during the operation of a MAC. The amount of hydrogen generated will depend upon the current applied to the cell. The process of hydrogen generation from the electrolysis of the electrolyte is also governed by mass transfer, growth of hydrogen bubbles, and removal of hydrogen from the MAC. Riegel [41] discusses the mechanisms for hydrogen removal as (1) hydrogen dissolved in the electrolyte solution is removed by diffusion and convection and (2) gas bubbles are

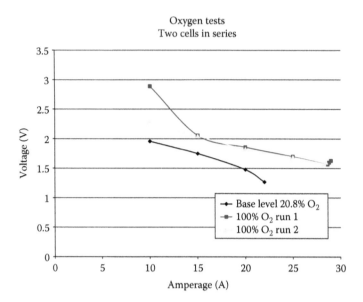

FIGURE 4.17   Effect of oxygen enrichment using two cells (ambient temperature).

transported by a two-phase flow. Thus, hydrogen gas will be diluted volumetrically. The use of a hydrogen inhibitor (HI) aids in reducing the amount of hydrogen generated at the anode.

The driving force for the gas transfer is the difference in partial pressure of gasses between the bubbles stuck in the air diffuse cathode and the air outside. The bubbles are saturated with water vapors, and they are connected with outside air through the pores in the cathode. The diffusion of nitrogen and oxygen to the bubbles is caused by the difference in partial pressure, resulting in the increase of volume of the gas bubbles. As this new gas volume gets saturated with water vapors, the process will continue [42].

The effect of oxygen enrichment is shown in Figure 4.17. When the oxygen concentration is high, a higher load can be utilized with minimizing effect on the voltage.

4.3.1   GAS PHENOMENA

At low amperage (<1), very few gas bubbles are noticed; however, at higher amperages, tiny gas bubbles are quite noticeable in a steady stream. At low amperage, the electrolyte temperature is room temperature; however, at high amperage, the temperature can range from 60°C to 85°C.

When hot tap water is poured into a cell, bubbles are noticed immediately rising to the top of the cell. This may be explained by the fact that the driving force for the gas transfer is the difference in partial pressure of gasses between the bubbles stuck in the electrode and the air outside. The bubbles are saturated with water vapor and they are connected with the outside air through the pores in the GDE. The diffusion of nitrogen and oxygen to the bubbles, caused by the difference in partial pressure, results in the increase of volume of gas bubbles. As this new gas volume resaturates with water vapor, the process continues.

This may be the phenomena that is observed when the cell operates at high amperage (i.e., reaction is fast thus quickly drawing in oxygen), which leads to an increase in the temperature of the electrolyte (exothermic reaction). Air bubbles appear as the electrolyte warms up. Air was dissolved in the water when it was cold, but now that the water is hotter, it can no longer stay in solution. So it appears as bubbles (i.e., gases come out of solution). The solubility of gases in water decreases with rising temperature. The gas bubbles contain nitrogen, oxygen, hydrogen, and water vapor.

4.4 ELECTROLYTE

Research has been mainly based on saline-based electrolytes; however, other aqueous-based electrolytes have been investigated [43,44]. The electrolyte is a saline solution such as seawater, common salt (NaCl) dissolved in water, or water containing a special salt mixture as prepared by various companies involved in the development of the magnesium–air technology. Other types that should work are naturally occurring brines and alkali pond waters. Urine has also been demonstrated to work effectively in the MAC [12]. The choice of electrolytes is fairly flexible, but the power output will differ for each of these electrolytes. The amount of magnesium anode corrosion is dependent upon the salt content of the electrolyte (Figure 4.18). Thus, the current also depends upon the conductivity of the electrolyte. Salt is highly corrosive to magnesium, but it (chloride ion) is needed for the conductivity of the water-based electrolyte and activates the magnesium

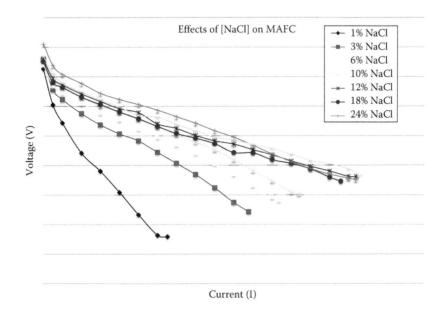

FIGURE 4.18 Effect of salt concentration (ambient temperature).

by breaking up the magnesium oxide/hydroxide layer on the anode. Thus, higher the salt concentration greater the corrosion rate. Additives may be used in the electrolyte to promote conductivity and help suppress hydrogen gas. However, an overly high salt concentration will lead to unwarranted and potential detrimental corrosion of the cathode, specifically the metal mesh (i.e., nickel). The pH of the saltwater-based electrolyte is neutral, and as the magnesium corrosion progresses resulting in production of magnesium hydroxide, the pH increases. In order to keep magnesium hydroxide in solution, the pH can be reduced by additives. Hydroxide saturation will occur upon water being consumed and will precipitate. Magnesium hydroxide and other precipitates can contribute to the reduction of cell efficiency and the fouling of cell components. The electrolyte management control is important to maintaining a "healthy" battery.

Seawater is a natural electrolyte that can be used in the magnesium–air battery. Seawater has an average salt concentration of 3.2% and can contain dissolved oxygen, which may aid in the corrosion of the magnesium anode and for reducing water to be the prevalent cathodic reaction. Since seawater can contain microorganisms, then microbially induced corrosion may be involved in material degradation, which may be beneficial [45]. Which type of microorganism is involved differs by the type of material and the environment. In environments with high oxygen content, aerobic bacteria are active, but when the oxygen level is lowered, these are replaced by anaerobic bacteria. Some bacteria are able to utilize hydrogen formed during the cathodic corrosion process. Test work [12] shows that seawater electrolyte is limited to loads of one ampere or less due to the salt concentration (see also Section 4.9.1). Use of this seawater electrolyte can lead to numerous devices and products that can make use of this natural electrolyte.

4.5   SEPARATORS

Separators are an integral part of batteries (i.e., lead–acid, zinc–air) but have not been reported as being used in the magnesium–air battery. The purpose of the separator is to prevent precipitate build up on the gas diffusion layer by repelling the precipitate but should be porous enough to allow air passage. The separator should be of sufficient quality to allow periodic cleaning/rinsing to remove any precipitate.

4.6   CURRENT COLLECTOR

The electrically conductive current collector is imbedded as part of the air diffusion cathode and provides the structure to this cathode but at the same time allows air/oxygen to diffuse through. The collector includes meshes or grids or foam made of such metals as aluminum, Inconel, stainless steel, copper, silver, titanium, or nickel. The conductive current collector may also be a graphite plate with channels provided in one face through which air/oxygen can flow. The collector should also have excellent corrosion resistance in aqueous saltwater electrolyte preventing forming of a nonconductive layer on the surface. The collector thickness, as well as the size and shape of openings in a mesh or foam, is important as they can impact the internal resistance in the cathode.

By combining the correct catalyst and current collector in conjunction with the type of electrolyte and anode, the performance of the air diffusion cathode can be impacted.

4.7 MANUFACTURING

The fabrication of components is extremely important in the development of the magnesium–air battery and will be dependent upon the device/product power requirements and usage. The design and choice of materials used in the manufacture of magnesium–air battery products are very important with regard to costs and environmental impact. Standardization of components (or modular components) can help reduce manufacturing and product development costs. Failure mode analysis of each component is important for product integrity and reliability.

The goal of all magnesium–air batteries and related products is about costs and creating a consumer accepted product.

4.7.1 COSTS

- The major cost is the air diffusion cathode that can be up to 1/3 of the product manufactured cost depending upon the catalyst and current collector used. Also, there are very few cathode manufacturers worldwide.
- The anode cost will depend upon the material/alloy(s) used and method of manufacturing (i.e., die cast vs. extrusion vs. foil). The anode shape and configuration can be rectangular, cylindrical, sheet, or button depending upon the end product. The cost of the anode will depend upon the world commodity price of magnesium.
- The electrolyte cost is relatively inexpensive assuming a saltwater solution.
- The major cost for injection-moldable parts is the tool(s) itself; however, the plastic medium used will depend upon how much recycled plastic is used and the type of plastic.

The basic components of a product built using the MAC (or any other aqueous metal–air power sources) in a basic 12 V configuration would include the following:

- Basic cell (12 cells in series configuration)
- Anode (12 anodes with at least 80 Ah capacity)
- Electrolyte (water based)
- Additives (added to electrolyte)
- Cathode (probably with cobalt catalyst)
- Tank to hold at least 3 L of electrolyte
- Pump (pump electrolyte through the cells that helps to maintain constant temperature and prevent precipitate build up on cathodes)
- Thermostat (monitor temperature of unit)
- Heat sink (absorb heat generated from voltage regulator)
- Fan (increase air flow through cathodes and cool unit)
- Receptor (12 V receptor for user device connection)

- Regulator board that regulates voltage, built in AC inverter, temperature shut-off, internal connectors for onboard power to fan, and 9 V battery recharger
- Casing (internal and external structure strong enough to withstand rugged usage)
- Packaging (labeling with brochures)

The manufacturing model would use worldwide injection mold manufacturers that will encompass approved manufacturing and assembly facilities. The objective of this is to ensure and maintain quality of product and reduce costs where appropriate. The finished product should have traceability throughout the process, including raw materials. After product manufacture, conformance testing is required that consists of the minimum number of tests and samples to assure conformance to requirements. This will be dependent upon each product specifications.

The manufacturing of the magnesium–air battery and components has some environmental impact concerns that must be addressed for the following reasons:

- Safeguarding of the health and safety of the employees and consumer and to protect the environment
  - Potential hazardous materials (material safety data sheets [MSDS] and Workplace Hazardous Material Information System [WHMIS])
    - The MSDS are published information on hazardous material in compliance with the WHMIS.
  - Potential electronic interference
    - All batteries create an electromotive force (emf) that must be taken into consideration when designing a battery device product. The electromotive force produced by primary (single-use) and secondary (rechargeable) cells is usually of the order of a few volts. The emf varies according to the size of the load and the state of exhaustion of the cell.
- Marketing and advertising information
  - Packaging and markings
    - The packaging, labeling, and information materials/brochures/manual(s) should be made using nontoxic materials and recycled materials (where economically possible). This process also will not cause nontoxic materials to be formed and/or dispersed in any form. The whole process will be environmentally friendly from start to finish.
- Manufacturing operations
  - The manufacturers must be ISO registered in their various countries.
- Codes and standards
  - Products using the metal air technology must undergo certification with respect to the following international codes and standards:
    - IEC-CSA 62282-5-1 (November 2007), which covers construction, markings, and testing for ac- and dc-type portable fuel cells, and IEC-CSA 62282-6-100, which covers micro–fuel cell power systems.

- Recyclability
  - Recycling of the magnesium–air battery is imperative to product manufacturing and selling. All materials used in the manufacturing of the battery must be either recyclable or disposable as well as any battery reaction products. The materials must be MSDS qualified.
- Disposability
  - If recycling is not possible, either due to location or other such means, then safe disposal of the magnesium–air battery is imperative to product manufacturing and selling. All reaction products from the battery usage must be either recyclable or disposable.
- Maintenance
  - Generally, a product that utilizes an aqueous metal–air power source will require periodic maintenance that should be conducted by the owner. This maintenance requires no specialized equipment nor is there any specialized training or service contract required. Maintenance may involve flushing the cells with water after each usage or when the anodes need replacing. Vinegar (acid) can be added to the wash water as it will dissolve the magnesium hydroxide (base) reaction product.

4.7.2 Cell Design

The magnesium–air battery would encompass an injection-moldable plastic and a cathode bus bar assembled and controlled in an injection mold tool for mass production. Since the cell technology lends itself to being scalable, then use of injection mold tools would be cost effective. Products using the MAC must also be designed for ease of assembly and disassembly that is paramount to recyclability. The smaller cell configurations can be recharged with magnesium by hand, whereas when power requirements involve numerous cells, then an automatic insertion of the anode material is needed. The MAC can be manufactured for stacking (Figures 4.19 and 4.20) or for standalone (Figure 4.21) depending on power requirements.

The important concept in constructing the cell is to minimize resistance where possible. This includes minimizing the gap between the anode and cathode (Figure 4.22), proper conductivity of the electrolyte, correct anode material, and precipitate control. The cell can be rectangular or cylindrical in shape and is scalable depending upon the power application. The cell is typically bifunctional where two GDEs are combined in parallel configuration.

4.7.3 Anodes

The anode is "sandwiched" between the cathodes with particular attention paid to the "gap" between the anode and cathode.

From test work and literature review, the chemistry, especially trace element content, is important in the development of the anode for the MAFC. It is imperative that the correct trace element chemistry be initiated. The anode chemistry should be assessed as to any impacts it will have on the environment (i.e., ecomaterial assessment).

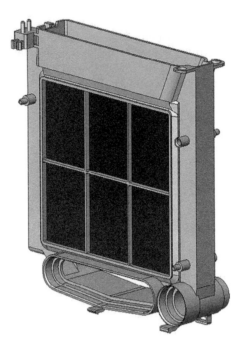

FIGURE 4.19   Moldable stacking magnesium–air cell.

FIGURE 4.20   Magnesium anode inserted above the cell.

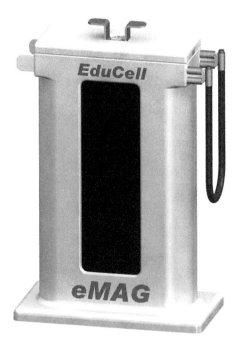

FIGURE 4.21   Stand-alone magnesium–air cell.

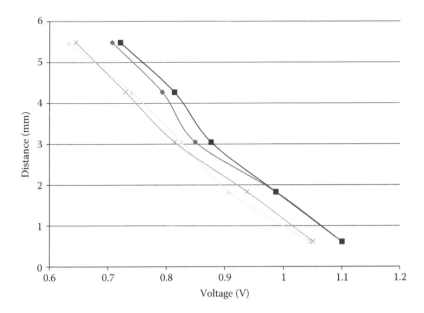

FIGURE 4.22   Effect of distance between anode and cathode on voltage over 4 runs, 5 A discharge. This gap will increase as the anode corrodes, thus increasing the resistance.

From test work, it is imperative that the correct fabrication process be initiated. The magnesium anode commercial fabrication methods/process to date are

- Die casting
- Thixomolding
- Metal injection molding (powder injection molding)
- Rheocasting (slurry on demand)
- Thixocasting (billet reheating)
- Metal powder (powder metallurgy—cold pressing and metal spray)
- Metal stamping (sheets)
- Foam (still in R&D)
- Extrusion

4.7.4   CATHODE

Neburchilov et al. [46] reviewed the compositions, design, and methods of fabrication of air cathodes for alkali zinc–air fuel cells, some of which may have some commercial use for the magnesium–air battery.

The cell itself is composed of two air-diffusion cathodes that are connected in parallel configuration and when stacked form a series configuration.

The current collector material is generally purchased in rolls and should be heat-treated before cathode assembly. This heat treatment will allow easier handling of the rolled material and also to remove any substance that may be coating the metal.

In conjunction with the current collector is the bus bar. The bus bar is a strip of metal (copper, aluminum) that conducts the current from the current collector to the electrical connectors of the cell.

Another aspect of the current collector to note is the orientation of the mesh openings as the bus bar must be joined (i.e., welded) to the current collector parallel to the long axis of the mesh openings. From practical experience, the electrical resistance is minimized when the bus bar is joined parallel to the long axis of the mesh openings.

4.7.5   ELECTROLYTE

There is no particular fabrication process to creating the aqueous electrolyte except that the correct amount of salt and additives are combined and inserted in an air-sealed container. Another method would be to combine the salt and additives into a tablet or disk shape. The user can use any source of "water" media (i.e., tap water, seawater, lake water).

4.8   APPLICATIONS

The fundamental properties of magnesium–air batteries that make these batteries important for commercial applications are (1) high energy density; (2) long, almost indefinite life in a dormant stage without impacting operational performance; (3) immediate activation when with the addition of electrolyte; and (4) high intrinsic safety of the system. The technology is best suited to consumers that require power on demand and who are willing to manage the system by replacing nontoxic anodes and electrolyte when required.

The magnesium–air technology as an alternative power source is generally limited up to 5 kW, though no commercial product has been manufactured to date beyond 100 W. The ability to penetrate the power markets with the magnesium–air technology is huge, but there are essentially no commercial MAC power products in the market at present. There are numerous theoretical applications, but these applications must be transformed into commercially acceptable cost-effective products. This technology will be new to the consumer and as such must be introduced on a "first to market basis," which will include some education and mass marketing. The variable cost to all products is the magnesium, which can vary with world market conditions. There are aqueous batteries [47] in the present market but that have an insignificant market share to date.

The consumer oriented products should be

- Environmentally safe
- Nontoxic
- Recyclable
- Indefinite storage
- Easily transportable
- Consumer friendly
- Cost competitive with other products

Magnesium air technology will be most applicable to

- Value-added power device for commercial applications
- Power device for emergency and disaster situations
- Power device for data loggers and battery chargers
- Camping and RV products
- Education kits and power toys
- Backup power
- Marine (surface and underwater applications)

Advantages of the aqueous magnesium–air battery technology include

- Indefinite shelf life (in a dry state)
- Nontoxic electrolyte
- Nonhazardous disposal materials
- Nontoxic fuel (anode)
- Nontoxic cathode
- No greenhouse gas emissions
- Fuel flexibility
- Electrolyte flexibility (note: no other battery uses seawater)
- Quiet
- No short circuit degradation
- Operating temperature range from −20°C to 60°C, depending upon electrolyte composition (based on test work to date)
- Safely transported in dry state

- Flexible storage temperatures in dry state
- Flexibility of manufacturing processes
- No catastrophic battery degradation when one or two cells fail in multiple cell configuration
- Low power density
- High energy density
- Magnesium stability in air and aqueous electrolytes
- Magnesium anodes not plagued by dendritic formation

4.9 CHALLENGES AND PERSPECTIVES

The magnesium–air technology has been limited in development due to several technological barriers, which have a profound impact upon the commercialization of the technology.

Two dominant disadvantages to the aqueous magnesium–air technology include the fact that the anode produces hydrogen as a parasitic reaction, where magnesium is consumed and hydrogen evolves with no electricity produced. Also, the product of the anode, $Mg(OH)_2$, will precipitate once the electrolyte supersaturates. These two technical challenges together with air diffusion cathode costs are likely the reasons that magnesium–air systems have not been widely commercialized by a battery company.

4.9.1 CHALLENGES: DEGRADATION MECHANISMS AND MITIGATION STRATEGIES

4.9.1.1 Hydrogen Problem

Hydrogen gas is generated by the spontaneous reaction of magnesium with water and magnesium metal is thermodynamically unstable in water with respect to the formation of magnesium hydroxide and hydrogen. This reaction is exothermic, which consumes magnesium. Low electrolyte pH may also enhance hydrogen evolution; thus it is desirable to have an electrolyte pH close to neutral. The parasitic reaction of magnesium with water is shown in Equation 4.4, which consumes magnesium and produces hydrogen without producing electricity.

The sensitivity of hydrogen evolution to pH can be explained by the following reactions:

Anode reaction

$$Mg - 2e^- + 2H_2O = Mg(OH)_2 + 2H^+ \tag{4.9}$$

Parasitic cathode reaction on anode

$$Mg + 2e^- + 2H^+ = MgH_2 \tag{4.10}$$

$$MgH_2 = Mg + H_2 \tag{4.11}$$

Overall parasitic reaction

$$Mg + 2H_2O = Mg(OH)_2 + H_2 \tag{4.12}$$

These reactions suggest that the way to stop parasitic decay of Mg by reaction (4.10) is by removing H^+ species with proton scavengers.

Hydrogen evolution is a secondary reaction that occurs on the anode. The MAC performance improves when hydrogen evolution is diminished, resulting in higher MUE. The minimization of hydrogen evolution that accompanies the magnesium electrodissolution is an important issue for the successful development of the MAC. Thus, an HI additive(s) needs to be used in the MAC, which inhibits hydrogen evolution on the anode surface. This can lead to deterioration of the anode surface.

The influence of a HI on the anodic polarization of magnesium is an essential part in the control of hydrogen generation in the magnesium–air–water interactions.

The reduction and management of hydrogen is important for several reasons:

- Reduce internal pressure in the cell and reduce water electrolysis.
- Reduce and control hydrogen from reaction in order to achieve greater energy density. This can be accomplished by reducing hydrogen bubbles and reduction of magnesium hydroxide precipitate, both of which add to the internal resistance of the cell.
- Reduce hydrogen gas bubbles and thereby reduce volume of cell. There is a measurable increase in cell volume in order to accommodate the gas bubbles.
- Increase the MUE of the MAC.
- Reduce the unlikely risk of explosion. The MAC will not function for long periods in a sealed space because the system consumes oxygen, and oxygen must be available to operate the cell. The system would become oxygen starved and cease operation before sufficient buildup of hydrogen would create a safety hazard.

4.9.1.2   Magnesium Utilization Efficiency

Various papers [48–51] have been published regarding the efficiency of magnesium and magnesium alloy materials.

The MUE of the MAC is measured via the amount of electric charge passed (i.e., electricity generated) over the weight loss of the magnesium anode and is referred to as the MUE.

$$MUE\% = I \times Ael \times t \times Mgw \times 100$$

$$2 \times F \times m \times at$$

where
  I is the current (A)
  Ael is the anode geometric area (m^2)
  t is the time (s)
  Mgw is the Mg atomic weight (24.3)
  F is the Faraday's number (96485.3)
  m is the weight loss (g)
  at is the anode type
  that is, anode type: AZ31 = 0.96; AZ61 = 0.93

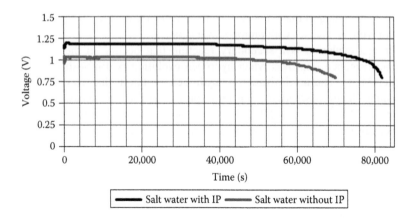

FIGURE 4.23 Effect of HI additive in electrolyte (1 cell at 5 A discharge, anode type, AM60B; active anode face = 99 cm²; active cathode face = 108 cm²). Salt concentration 10%. (From Vizon Scitec, Inc., July 2003.)

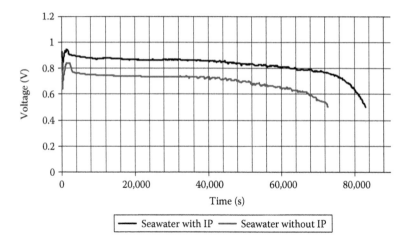

FIGURE 4.24 Effect of HI additive in seawater 3.2% salt concentration (From Vizon Scitec, Inc., July 2003) (1 cell, 5 A discharge; anode type, AM60B; active anode face = 99 cm²; active cathode face = 108 cm²).

An HI additive can be used to help suppress hydrogen gas (and increase the MUE of the anode) that is naturally formed at the anode. It is added to the electrolyte and should be nontoxic, water soluble, and nonpolluting to the environment. Figures 4.23 and 4.24 show the effect of an HI added to the saltwater-based electrolyte. With the HI, the MUE of anode in the MAC is 70%–75%, while without the HI the MUE of anode is 45%–55%.

4.9.1.3 Electrolyte

Another limitation is the solubility of the anode reaction product(s), (i.e., magnesium hydroxide, $Mg(OH)_2$). Magnesium hydroxide is soluble in water, but as it is continuously produced during operation, it will saturate and eventually precipitate from the electrolyte solution. Therefore, the magnesium–air battery will require user management to maintain its optimum performance.

The water portion of the electrolyte is consumed and needs to be replaced over a period of time, which depends on the load of the power device. Magnesium discharging into the electrolyte combines with hydroxide and other ions (i.e., sodium, chloride) to form various magnesium reaction products as identified by x-ray diffraction. The electrolyte needs to be replaced before reaching the solubility limits of the magnesium reaction products. There will be an increase in cell resistance with increasing mass of reaction products, resulting in an MUE decrease. These reaction products can also adhere to the cathode, thereby plugging the pores through which air passes, which will cause a reduction in the amount of oxygen passing through the air diffusion cathodes and thus a potential decrease in the MUE.

Pumping of the electrolyte in certain commercial products will help to alleviate the stagnation problem of stand-alone cells. It is important to ascertain the volume of electrolyte that is required for the needed power and run time, which impacts the size of the cell and/or electrolyte tank. It is also important to note in the design of a product that utilizes a common electrolyte tank to the cells that electrolyte shorting can occur between cells.

Another impact of $Mg(OH)_2$ precipitation is that high ionic conductivity aqueous electrolytes such as KOH cannot be used. Therefore, electrolyte choice is limited to salts and/or additives that maintain the pH as close to neutral as possible. This limits the choice of electrolytes at the cost of lower power density. Instead of increasing the ionic conductivity of the electrolyte, one can compensate for low power density by increasing the number of cells in the system in order to meet the power needs of each application.

4.9.1.4 Precipitate

A significant buildup of precipitate on the anode and cathode surfaces seriously degrades the performance of the cell and service lifetime of the air cathode. Precipitate can block air flow through the cathode and create a short circuit between the cathode and anode. Precipitate on the anode can also block electrolyte contact with the magnesium anode, inhibiting conductivity and slowing the anode consumption rate. In this case, the magnesium efficiency could be inaccurate as the precipitate forms a barrier between the electrolyte and anode surface.

Precipitate buildup will reach a critical stage as water is consumed, and the electrolyte becomes saturated and reaction product(s) starts to precipitate as a solid. The electrolyte can be refreshed with the addition of water and/or simply changed with new electrolyte. The magic "elixir" would be an additive that binds magnesium ion from combining with the hydroxide ion to forming magnesium hydroxide and/or use a flocculent to control precipitates.

4.9.1.5 Cathode

One of the limiting factors is the difficulty in developing cost-effective, simple, reliable cathode structures, which deliver high performance, optimize cathode catalyst recipe specifications, optimize cathode mass transport architecture structure, and allow economic manufacturing processes. For instance, current commercially developed air diffusion cathodes typically have problems of high cost, high internal electrical resistance, and corrosion to metal current collector layer in the alkaline or neutral electrolyte environments. Most of the developed air diffusion cathodes for metal–air batteries are made for the alkaline electrolyte environment, which may not be suitable for neutral or salt (i.e., sodium chloride) electrolyte environments. New one layer cathodes are being developed, which can be cost effective on large-scale manufacturing volume. The cost of the catalyst, whether it is cobalt or manganese based, has a direct bearing on the cathode cost. Typically, the cathode cost is approximately 1/3 the cost of a MAC. The choice of catalyst should be made on the design and power requirements of the cell. If the current is less than 1 A, then a manganese catalyst should be used.

In the design of larger systems, a fan would provide constant air flow, thereby eliminating any air stagnation problem and improving oxygen reduction reaction.

4.9.2 PERSPECTIVES

The development progression is to develop a mechanically and automatically rechargeable MAC using

- A cost-effective metal anode
- Cost-effective cathode materials and structures as required to achieve reliability and efficiency goals
- A conductive electrolyte

This is done to establish the economic competitiveness of the technology, and to develop and integrate systems utilizing the MAC.

Another key challenge is market acceptance. The success of this technology driven products hinges on whether or not consumers will find the benefits of these technology products to be sufficiently attractive at the price. Because this technology is fundamentally so different from any other power source available, educating potential buyers will be the key promotional challenge.

As with all potential applications, there needs to be a market analysis for the various applications that would include the following:

- Existing metal–air market scenario
- Existing magnesium–air market scenario
- Market metal–air forecasts
  - Market magnesium–air forecasts
- Market drivers
  - Market restraints
  - Pricing analysis
  - Market trends
    - North America
    - Europe
    - Asia
    - Rest of the world

- Identification of metal–air markets (potential applications/products)
- Identification of magnesium–air markets (potential applications/products)
- Identification of industry challenges
- Competitive technologies
  - Advantages
  - Disadvantages
- Magnesium market and magnesium producers
  - Forecast/supply/demand

4.10 SUMMARY

Metal–air technologies have some limitations and challenges to overcome but will have an impact on alternative energy sources. Though there exists the high theoretical energy density for metal–air batteries, it is the battery efficiency that is the major goal. One battery technology is not the magic goal, but each battery technology needs to be reviewed on its own merit(s) and commercial products developed on that basis.

Magnesium–air battery technology has substantially evolved since the 1960s whereby commercial applications will become accepted. The production of the nonaqueous rechargeable magnesium–air battery is still challenging, but it has the potential to replace other batteries specifically for electrically powered vehicles.

The aqueous magnesium–air battery has had some commercialized production but is limited to novelty products. A major commercialization for this technology is the backup power product market.

From review of papers regarding magnesium–air batteries, none really delve into the commercialization of the technology with respect to consumer products.

Further research and development of magnesium–air batteries must not just involve chemists/electrochemists and materials engineers/metallurgists but must also include people versed in product development, manufacturing, and marketing.

There exists multiple opportunities for magnesium–air batteries.

REFERENCES

1. WP ST 9-Metal/Air, 2002, Institute for Environment and Sustainability Renewable Energies Unit, Ispra Site, Italy, Storage Technology Report.
2. Linden, D. and Reddy, T.B. (eds.), 2002, *Handbook of Batteries*, 3rd edn., McGraw-Hill, New York.
3. Kaisheva, A., 2005, Metal-air batteries: Research, development, application, *Proceedings of the International Workshop: Portable and Emergency Energy Sources—From Materials to Systems*, Primorsko, Bulgaria, September 2005.
4. Aral, H. and Hayashi, M., 2009, Secondary batteries—Metal-air systems, *Encyclopedia of Electrochemical Power Sources*, 4, 347–365.
5. Egashira, M., 2009, Iron-air (secondary and primary), *Encyclopedia of Electrochemical Power Sources*, 4, 372–375.
6. Lee, J.-S., Kim, S.T., Cao, R., Choi, N.-S., Liu, M., Lee, K.T., and Cho, J., 2011, Metal-air batteries with high energy density: Li–Air versus Zn–Air, *Advanced Energy Metals*, 1(1), 34–50.
7. Energy Storage, 2002, March/April, Refocus, Elsevier Advanced Technology Article.

8. Tuck, C.D.S. (ed.), *Modern Battery Technology*, Ellis Horwood Publishers, pp. 487–502.

9. Fitzpatrick, N. and Scamans, G., 1986, Aluminum is a fuel for tomorrow, *New Scientist* July 17, 1986, pp. 34–39.

10. Visco, S.J., Nimon, E., and De Jonghe, L.C., 2009, Lithium-air, *Encyclopedia of Electrochemical Power Sources*, 4, 378–383.

11. Downing, B.W., 2012, Metal air technology, Chapter 6, in *Electrochemical Technologies for Energy Storage and Conversion*, volume 1, edited by R.-S. Liu, L. Zhang, X. Sun, H. Liu, and J. Zhang, Wiley-VCH Verlag GmbH & Co, pp. 239–277.

12. Downing, B.W., 2003 to 2008, MagPower Systems, Inc. MagPower unpublished reports.

13. Hamlen, R., Jerabek, E.C., Ruzzo, J.C., and Siwek, E.G., 1969, Anodes for refuelable magnesium-air batteries, *Journal of the Electrochemical Society*, 116(11), 1588–1592.

14. Hasvold, O. et al., 1997, Sea-water battery for subsea control systems, *Journal of Power Sources*, 65, 253–261.

15. Sathyanarayana, S. and Munichandraiah, N., 1981, A new magnesium-air cell for long life application, *Journal of Applied Electrochemistry*, 11, 33–39.

16. Wilcock, W. and Kauffman, P.C., 1997, Development of a seawater battery for deepwater applications, *Journal of Power Sources*, 66, 71–75.

17. Kent, C.E. and Carson, W.N., 1966, *Magnesium-Air Cells*, Technical Information Series, Research and Development Center, General Electric.

18. Zhang, T., Tao, Z., and Chen, J., 2014, Magnesium-air batteries: From principle to application, *Materials Horizons, The Royal Society of Chemistry*, 1, 196–206.

19. Kim, H., Jeong, G., Kim, Y.-U., Kim, J.-H., Park, C.-M., and Sohn, H.-J., 2013, Metallic anodes for next generation secondary batteries, *Chemical Society Reviews*, 42, 9011.

20. Rahman, Md.A., Wang, X., and Wen, C., 2013, High energy density metal-air batteries: A review, *Journal of the Electrochemical Society*, 160(10), A1759–A1771.

21. Winther-Jensen, B., Gaadingwe, M., Macfarlane, D.R., and Forsyth, M., 2008, Control of magnesium interfacial reactions in aqueous electrolytes towards a biocompatible battery, *Electrochimica Acta*, 53, 5881–5884.

22. Abraham, K.M., 2008, A brief history of non-aqueous metal-air batteries, *ECS Transactions* (the electrochemical society), 3(42), 67–71.

23. Aurbach, D., Suresh, G.S., Mizrahi, O., Mitelman, A., Levi, E., and Levi, M.D., 2008, New rechargeable magnesium battery systems, *213th ECS Meeting*, Abstract #289, The Electrochemical Society.

24. Muldoon, J., Bucur, C., Oliver, A.G., Sugimoto, T., Matsui, M., Kim, H.S., Allred, G.D., Zajicek, J., and Koyani, Y., 2012, Electrolyte roadblocks to a magnesium rechargeable battery, *Energy Environmental Science*, 5, 5941–5950.

25. Mohtadi, R. and Mizuno, F., 2014, Magnesium batteries: Current state of the art, issues and future perspectives, *Beilstein Journal of Nanotechnology*, 5, 1291–1311.

26. Yoo, H.D., Shterenberg, I., Gofer, Y., Gershinsky, G., Pour, N., and Aurbach, D., 2013, Mg rechargeable batteries: An on-going challenge, *Energy and Environmental Science*, 6, 2265–2279.

27. Song, G.L. and Atrens, A., 1999, Corrosion mechanisms of magnesium alloys, *Advanced Engineering Materials*, 1(1), 1438–1656.

28. Weber, C.R., Knörnschild, G., and Dick, L.F.P., 2003, The negative-difference effect during the localized corrosion of magnesium and of the AZ91hp alloy, *Journal of the Brazilian Chemical Society*, 14(4), 584–593.

29. Atrens, A. and Dietzel, W., 2007, The negative difference effect and unipositive Mg^+, *Advanced Engineering Materials*, 9(4), 292–297.

30. Liu, L.J. and Schlesinger, M., 2009, Corrosion of magnesium and its alloys, *Corrosion Science*, 51, 1733–1737.

31. Ficara, P. et al., 1998, Magnola: A novel commercial process for the primary production of magnesium, *CIM Bulletin*, 91(1019), 75–80.

32. Bakke, P., Svalestuen, J.M., and Albright, D., 2002, Metal quality—The effects on die castors and end users, Society of Automotive Engineers, Paper 2002-01-0078.
33. Shaw, B.A., 2003, Corrosion resistance of magnesium alloys, *ASM Handbook, Vol. 13A Corrosion Fundamentals, Testing and Protection*, ASM Handbook.
34. Parthiban, G.T., Palaniswamy, N., and Sivan, V., 2009, Effect of manganese addition on anode characteristics of electrolytic magnesium, *Anti-Corrosion Methods and Materials*, 56/2, 79–83.
35. Zanotto, F., 2009, Corrosion behaviour of the AZ31 magnesium alloy and surface treatments for its corrosion protection. PhD thesis, Università degli Studi di Ferrara, Corrosion Study Centre, Italy.
36. Eliezer, D. and Alves, H., 2002, *Corrosion and Oxidation of Magnesium Alloys*, *Handbook of Materials Selection*, John Wiley & Sons, pp. 267–291.
37. Liu, M., Uggowitzer, P.J., Nagasekhar, A.V., Schmutz, E., Mark, S., and Guang-Ling, A.A., 2009, Calculated phase diagrams and the corrosion of die-cast Mg-Al alloys, *Corrosion Science*, 51, 602–619.
38. Avedesian, M.M. and Baker, H. (editors), 2009, *Magnesium and Magnesium Alloys*, *ASM Specialty Handbook*, ASM International.
39. Gonzalez-Torreira, M., Fones, A., and Davenport, A.J., 2003, Passivation and dissolution of magnesium, *Journal of Corrosion Science and Engineering*, 6, Paper C034.
40. Tang, Z., Xie, Z., Staite, M., Ghosh, D., Downing, B., and Jung, J., 2005, Anode material and fabrication development for magnesium-air fuel cell, COM 2005—*44th Annual Conference of Metallurgists*, Calgary, Alberta, Canada, Paper 43.4.
41. Riegel, H., Mitrovic, J., and Stephan, K., 1998, Role of mass transfer on hydrogen evolution in aqueous media, *Journal of Applied Electrochemistry*, 28, 10–17.
42. Halliop, W., 2001, Personal communication, email, May 13, 2001.
43. Guadarrama-Munoz, F., Mendoza-Flores, J., Duran-Romero, R., and Genesca, J., 2006, Electrochemical study on magnesium anodes in NaCl and CaSO4–Mg(OH)$_2$ aqueous solutions, *Electrochimica Acta*, 51, 1820–1830.
44. Vuorilehto, K., 2003, An environmentally friendly water-activated manganese dioxide battery, *Journal of Applied Electrochemistry*, 33, 15–21.
45. Jung, S., Yoon, M.-H., Lee, S.-M., Oh, S.-E., Kang, H., and Yang, J.-K., 2014, Power generation and anode bacterial compositions of sediment fuel cells differing in anode materials and carbon sources, *International Journal of Electrochemical Science*, 9, 315–326.
46. Neburchilov, V., Wang, H., Martin, J.J., and Qu, W., 2010, A review on air cathodes for zinc–air fuel cells, *Journal of Power Sources*, 195, 1271–1291.
47. Sammoura, F., Lee, K.B., and Lin, L., 2004, Water-activated disposable and long shelf life microbatteries, *Sensors and Actuators A*, 111, 79–86.
48. Gummow, R.A., 2004, *Performance efficiency of High-Potential Mg Anodes for Cathodically Protecting Iron Water Mains*, CORRENG Consulting Service, Inc., May 2004 Materials Performance.
49. Yunovich, M., 2004, *Performance of High Potential Magnesium Anodes: Factors Affecting Efficiency*, CC Technologies Laboratories, Inc.
50. Campillo, B., Rodriguez, C., Juarez-Islas, J., Genesca, J., and Martinez, L., 1996, *An Improvement of the Anodic Efficiency of Commercial Mg Anodes*, Corrosion 96 conference, paper 201.
51. Di Gabriele, F. and Scantlebury, J. D., *Corrosion Behaviour of Magnesium Sacrificial Anodes in Tap Water*, 2003, Corrosion and Protection Centre.

5 Metal–Sulfur Batteries
Fundamentals and Applications—A Perspective

L.C. De Jonghe, E.S. Nimon,
P.N. Ross, and S.J. Visco

CONTENTS

5.1   INTRODUCTION

Lithium–sulfur (Li/S) battery chemistry has a long history. Early work and patents, including lithium anodes and sulfur-type cathodes, date back at least 50 years [1], owing to the potentially high energy densities of this system. Despite its long development history, rechargeable Li/S batteries have yet to reach performances suitable for competitive commercial markets. A measure of the growth in interest may be seen in the increasing number of peer-reviewed publications over the last few years [2] (Figure 5.1). To date, numerous vigorous efforts aimed at capitalizing on the promise of the unsurpassed gravimetric energy densities are continuing, and progress is being made, albeit slowly.

We review some of the recent developments of rechargeable Li/S batteries that address several interrelated challenges: capacity decline on cycling, reactions of the lithium metal anode with cathode materials, and capacity limitation at increasing

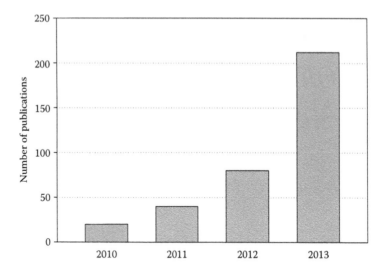

FIGURE 5.1   Annual peer-reviewed publications on lithium/sulfur systems, not including conference presentations.

current densities. We also relate some of the potential challenges in the development of rechargeable Li/S batteries to the earlier relevant developments of sodium/sulfur (Na/S) batteries.

5.2   DEVELOPING RECHARGEABLE Li/S BATTERIES

Two common issues in Li/S batteries stem from reactions between Li and the polysulfides of the cathode. Different types of Li–S interactions occur whether the polysulfide species are solvated [3] or immobilized [4]. An issue, earlier described by Dey [5], involved the possibility of soluble polysulfides passivating the lithium metal anode, the effect of which could be delayed by the use of an ion exchange membrane as a separator. Later, a polysulfide redox shuttle was identified that, on the one hand, allowed for an overcharge process, but, on the other hand, contributed to capacity fade on cycling. A first description of the shuttle mechanism was incorporated in the patent by Visco et al. [6], and later discussed in more detail by Mikhaylik and Akridge [7]. The mechanism is one whereby oxidized sulfur species, produced, for example, on charging, diffuse to the lithium anode where they are reduced by direct chemical reaction on the lithium metal anode and are subsequently transported back to the cathode to be reoxidized.

Significant efforts have been directed at suppressing or eliminating the polysulfide redox shuttle that leads to high rates of self-discharge due to direct chemical reaction of polysulfides with unprotected lithium metal. One early approach involved reducing the solubility of polysulfides in the nonaqueous electrolyte by the addition of nonpolar solvents like toluene [8], but this led to unacceptable cell performance due to poor ionic conductivity of the electrolyte and premature polarization of the sulfur electrode due to precipitation of insoluble polysulfides. More recently, several groups have reported success with capillary confinement [9], adsorption on various carbon nanostructures [10], or encapsulation schemes such as proposed by Cui et al. [11]. Similar ideas were explored by scientists at Moltech (now Sion Power) more than a decade ago [12]. Moltech researchers synthesized a variety of conducting polymers with covalently bonded, immobilized polysulfide (or organosulfur) species. The approach involved bonding polysulfides to a conducting polymer backbone that mediated the oxidation/reduction reaction so that polysulfide species would be immobilized. Unfortunately, there is a basic problem with this approach. As can be seen in Figure 5.2, in the fully oxidized state, the polysulfide chain is indeed immobilized by covalent attachment to the polymer backbone, but after a simple two-electron transfer, the polysulfide chain is cut loose and free to diffuse into the electrolyte, as was evident from the cycling data that looked nearly identical to conventional nonaqueous Li–S charge/discharge curves.

Over the past decade or so, there have been many claims of success in taming the polysulfide shuttle and improving the cycle life of the sulfur electrode in Li–S cells. However, quite often the voltage–time curves look very similar to those for conventional Li–S cells with dissolved polysulfides. In particular, the discharge curves have a characteristic region, which is typical for reduction of long-chain polysulfides. In many of these reports, it is difficult (if not impossible) to determine the amount of electrolyte used in the laboratory experiments. Notably, excess electrolyte has a large and positive effect on cycling. In a recent paper from the Fraunhofer Institute for Chemical Technology [13], scientists characterized the performance of Li–S pouch cells over a broad range of electrolyte-to-sulfur ratios (E/S in mL/g) and claimed that commercially interesting cells could only be achieved with E/S ratios of three or less. Not unexpectedly, the Fraunhofer scientists observed excellent sulfur electrode capacity ($mAh/gram_{electrode}$)* on charge and discharge for high E/S ratios, but vanishingly small capacity at $E/S = 3$, the threshold for commercially viable Li–S systems. Until the aforementioned nanoconfinement approaches are tested in Li–S cells with well-defined E/S ratios as in the Fraunhofer experiments, the benefit of these approaches for commercial systems remains unproven. In contrast, the switch from nonaqueous to aqueous electrolytes does yield a profound and verifiable benefit (see Section 5.11). In the PolyPlus approach, highly concentrated aqueous polysulfide solutions are prepared by the reaction of the appropriate amounts of Li_2S, S, and water. Since aqueous polysulfides are in solution rather than loaded into a carbon matrix, direct comparison to the Fraunhofer study requires calculation of the total mass of sulfur in solution, added to the mass of the carbon matrix to yield $gram_{electrode}$ values. PolyPlus sulfur molarity values can similarly be converted to E/S ratios ($E/S = 3$ is approximately equal to 10 MS). Aqueous Li–S cells perform remarkably better than the nonaqueous Li–S cells reported by Fraunhofer. Aqueous Li–S cells tested at PolyPlus demonstrate discharge capacities of >430 $mAh/gram_{electrode}$ at E/S ratios of 2.6:1 (corresponding to 12 MS), while the nonaqueous Li–S cells achieved only 50 $mAh/gram_{electrode}$ at E/S ratios of 3:1.

---

* In Fraunhofer experiments, electrode weight includes carbon, binder, and loaded sulfur.

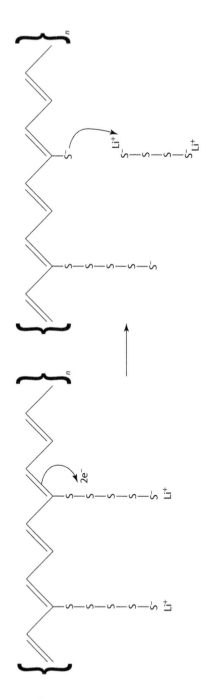

FIGURE 5.2 Polysulfides immobilized by covalent bonding to a conducting polymer backbone, in this case polyacetylene.

Much effort has gone into fully protecting the Li-metal anodes from the cathode chemistry. Polymeric membranes have not been fully adequate, as may be concluded from the recent review by Zhao et al. [14]. The main reason is that solvents, necessary for the operation of the sulfur cathode [15], invariably penetrate polymeric membranes, carrying the sulfur species with them. Chemical protection of the metallic lithium anode has also been only partially successful. For example, addition of lithium nitrate to the electrolyte, reported by Mikhaylik [16], has shown promise but provides only a temporary benefit [17] since nitrate is consumed in subsequent cycling due to its reaction with the lithium surface after partial breakdown of an SEI. Therefore, extended cycling requires the complete chemical protection of the lithium metal by a dense, single-ion conducting glass or ceramic. The LiPON glasses developed by Bates et al. [18] and the lithium–aluminum titanium phosphate (LATP) reported by Aono et al. [19] and by Fu [20], as well as the multilayer approach by Mikhaylik et al. [21], have been most promising to date. In addition, ceramic electrolyte compositions with the garnet structure have been reported by Weppner and associates [22] to be stable in contact with lithium and possibly also with aqueous electrolytes [23]. The use of electronically insulating liquid, polymeric, or solid-state interlayers between the lithium metal and the LATP separator can avoid the destructive chemical reduction of the LATP that results from direct contact with lithium metal [24]. In a number of reports, battery cells having protected lithium electrode structures, identical to those described in [24], have been termed "hybrid electrolyte" cells [25,26].

The necessity of fully protecting the lithium metal electrode in the Li/S system follows from the observation that attempts to protect the lithium electrode from reacting with polysulfides using polymeric membranes, reactively formed solid electrolyte films, or by sulfur species immobilization have not been fully effective. We therefore emphasize here the use of lithium anodes that are protected by a solid-state ceramic, glass, or glass–ceramic lithium-ion-conducting membrane.

5.3 SOLID CERAMIC ELECTROLYTES

The first application of a ceramic ionic conductor in an alkali metal anode battery was formulated by Kummer and Weber in 1966 [27]. The ceramic electrolyte was a polycrystalline beta-alumina with unusually high sodium ion conductivity. Its crystal structure was known much earlier [28]. The remarkable stability of the sodium beta-aluminas had been long appreciated and they were used, for example, as liners for glass melting furnaces [29]. Early single crystals of sodium beta alumina were, in fact, extracted from such liner refractory bricks [30]. A recent paper by Wen et al. [31] reviews the status of the Na/S technology and suggests some further potentially beneficial developments, including alternative cell geometries [32], and lowering the operating temperature of the Na/S or related sodium systems [33].

The Li/S system has some similarity to the Na/S system, and thus the experience with Na/S may be of value to Li/S development. Two important differences, however, complicate the comparison. As follows from the phase diagram [34], Na/S must operate at about 300°C or above, in the liquid regime between S and Na_2S_3 [35], limiting the sulfur utilization to 33% [36]. A broad liquid-phase regime in Li/S [37] is present only above 860°C, so that a liquid Li_2S_x cathode approach is not practical for traction applications even if a chemically stable electrolyte were found. Moreover, high-purity lithium generated on charge is far more aggressive than sodium, and thus direct electrolyte/lithium contact is more problematic as has been widely appreciated. Just as for Na/S, the Li/S battery anode has to be metallic to avoid unduly lowering the cell voltages and compromising energy and power densities. As a result of these limitations, as well as due to the kinetic stability of metallic lithium to some aprotic solvents, development of rechargeable Li-metal anode batteries has revolved around room temperature and sub–room temperature applications where the lithium metal is solid. A further consideration is that the sulfur systems have open-circuit voltage between –2.1 and –2.4 V, so that, obviously, more Na or Li needs to be transported than is the case for the high-voltage solid oxide cathode systems, to achieve the same energy densities. Despite the differences between the two systems, some lessons from Na/S development are applicable to the Li/S system.

5.4 DENDRITE FORMATION IN SOLID-STATE ELECTROLYTES

A troublesome ceramic electrolyte failure degradation mechanism was found early on for the Na/S system and became the subject of much research. The failure mechanism involved the propagation of sodium-filled cracks through the ceramic, driven by continued plating of sodium into such cracks during the charge cycle. The mechanism was first described by Brennan [38] who used an internally pressurized Griffith crack model. The model was later refined by a number of workers employing fracture mechanics concepts, with the same qualitative results. The propagation of a crack, pressurized due to the need to extrude the internally deposited sodium, is proportional to the sodium viscosity and the fracture toughness of the ceramic electrolyte. A critical current density can thus be formulated with some assumptions of crack geometry and current focusing. This mechanism is very general and could also operate for ceramic Li-ion conducting electrolytes, such as lithium lanthanum zirconates (LLZs), in contact with metallic lithium. Thus, dendrites can penetrate ceramic electrolytes driven by this mechanism in spite of the fact that the Young's modulus of materials such as beta-alumina is many times that of lithium metal (~200 GPa vs. 4.9 GPa). This is in contrast to the work of Monroe and Newman [39–41], predicting that an electrolyte modulus on the order of 10 GPa might be sufficient for dendrite initiation suppression at Li polymeric-type electrolyte interfaces, based on a stability analysis of continuous sinusoidal perturbations of a lithium/polymer interface. The difference resides in the unavoidability of preexisting crack-like flaws in the surfaces of polycrystalline ceramics, whereas the Monroe–Newman model does not include such materials flaws or fracture mechanics concepts, since only viscoelastic electrolytes are considered. It is intriguing that extrusion-based dendrite initiation was also considered by Stark et al. [42] to be a factor for stability of lithium/polymer interfaces, when the SEI layer ruptures during electrodeposition. However, quantitative predictions are more difficult for this mechanism since little is known about the mechanical properties of SEI layers.

Below the melting point of sodium, the effective viscosity becomes the extrusion flow stress of the solid metal, generating much higher pressure in the crack. Consequently, the critical current density for crack initiation, as monitored with acoustic emission detection, drops dramatically by as much as a factor of 10^3 as the sodium transitions from liquid to solid, as was evident in the work of De Jonghe et al. [43]. Above the melting point of sodium, electrolyte dendritic penetration of the ceramic is initiated in the A/cm^2 range, whereas at room temperature failure initiation occurred around a few mA/cm^2. It is worth noting that at similar strain rates, the flow stress of solid lithium is higher by a factor of as much as 10 [44], compared to solid sodium, predicting a failure by lithium dendrite initiation in a ceramic electrolyte at current densities on the order of 100 μA/cm^2. It is interesting to note that dendrite penetration of LLZ ceramic electrolytes indeed was recently reported [45], further supporting the general nature of the Li-dendrite formation mechanism.

Since crack propagation depends on the fracture toughness, K_{IC}, of ceramic or glass solid electrolytes, strategies to increase K_{IC} are beneficial in allowing higher critical current densities. Such strategies include refinement of the grain size and the incorporation of transformation toughening zirconia dispersions [46], which eventually has led to highly durable ceramic electrolytes for the Na/S system. Similar strategies for Li-ion ceramic conductors, such as LATP [20] and LLZ [22], have not been explored, presumably due to the reaction of zirconia dispersion with the matrix material. However, a correlation should be expected between the fracture toughness and critical current densities for dendrite initiation in both the beta-alumina-type solid electrolytes [47,48] and the LLZ ceramics, as well as other solid-state electrolytes.

It may be concluded that an important area of materials research is the development of high toughness ceramic electrolytes. In addition, since much of the failure initiation depends on the presence of preexisting surface flaws, which in ceramic electrolytes is closely related to the maximum grain size, developing highly conducting glass membranes would be equally important. The highest conductivities presently known are in the lithium–sulfide-based glass systems, with reported Li-ion conductivities over 10^{-4} S/cm [49]. Remarkably high conductivities of over 10^{-2} S/cm were reported for $Li_{10}GeP_2S_{12}$ by Kamaya et al. [50] and by Kuhn et al. [51]. Nagao et al. also examined the performance of $Li_2S–P_2S_5$ electrolytes in all-solid-state cell. Similar to the beta-alumina and LLZ solid electrolytes, lithium dendrites were found to intrude in these solid electrolytes via pores or interparticle boundaries at current densities above 1 mA/cm^2, while no dendrite were observed at 50 μA/cm^2 [52]. This again points to the universality of the dendrite formation mechanism for solid electrolytes in direct contact with the lithium anode. These electrolytes are presently under development for all-solid-state batteries. LIPON glasses, with room temperature Li-ion conductivities between 10^{-6}–10^{-7} S/cm [18], have also received much attention, but require supported thin-film electrolytes for battery applications other than microbatteries [53,54].

5.5 INTERFACE PORE FORMATION IN ALL-SOLID-STATE BATTERIES

An additional potential issue for all-solid-state batteries in which a lithium anode directly contacts the solid electrolyte is the formation of pores at the lithium/solid electrolyte interface during cycling, particularly during battery discharge. The basic mechanism for pore formation is analogous to that of oxidation of metals where an oxide scale grows by the outward diffusion of the metal ion to form oxide at the external surface of the oxide scale. This process would produce a supersaturation of lithium vacancies in the metal surface at the Li/solid electrolyte interface. Pores may then develop at the interface as a result of vacancy agglomeration, disrupting the anode/electrolyte contact. Pore formation is even possible in the interior of the lithium electrode, when vacancies that are not eliminated on the lithium surface diffuse into the metal bulk and agglomerate. The mechanism of pore formation has been extensively modeled, for example, by Bobeth et al. [55], for pore nucleation and growth during oxidation of pure metals such as aluminum and for metal alloys. It is intriguing to note that the problem of pore formation during oxidation is particularly severe for metal alloys, such as Li-Fe, or in the presence of sulfur contaminants. It is at present not clear whether this issue is avoidable, particularly when 0.5 or more mAhs of Li are cycled at high current densities when a high concentration of vacancies could be produced. A mechanism for interface vacancy annihilation has to be invoked, possibly involving sweeping of interface Li-terraces or Li-ledges, in a process akin to edge dislocation creep. In general, one may expect that very-high-purity lithium, such as that produced during the charge cycle, is less likely to be susceptible to problematic interface pore formation. This phenomenon has not yet been modeled for lithium anode/solid electrolyte interfaces, and invites further experimental and theoretical study.

5.6   SULFUR UTILIZATION

A critical factor for practical Na/S and Li/S rechargeable batteries is the ability to operate with the highest possible sulfur molarity at the highest possible utilization, so as to achieve competitive cathode energy densities. High concentrations of solid reaction products limit utilization and rates very severely since the redissolution of solid discharge products, such as Li_2S_2 or Na_2S_2, is often rate limiting. This is particularly the case with Li/S that operates at room temperature and below. Dilute Li/S systems have fewer problems in this regard [56] because, simply, there is less precipitate to dissolve in the available solvent. Extrapolations of rates and sulfur utilizations obtained on dilute systems or with shallow cycling, often characteristic of much of the research reported in the literature, are thus rarely, if ever, warranted.

The sulfur utilization is readily calculated if one considers the charge–discharge limits: when the fully charged state is Li_2S_x, then x corresponds to the total starting sulfur concentration in the cathode. The end of discharge is Li_2S_y so that, since S does not leave the cathode, the charged-to-discharged reaction can be written as

$$Li_2S_x + 2(x/y - 1) \, Li \rightarrow (x/y) \, Li_2S_y \qquad (5.1)$$

The fractional sulfur utilization, F_u, is then simply

$$F_u = (1/y - 1/x) \qquad (5.2)$$

It is instructive to represent the results graphically as in Figure 5.3.

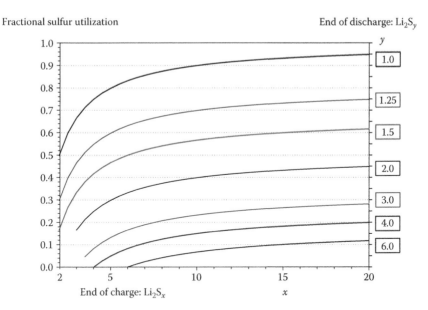

FIGURE 5.3 Maximum sulfur utilization for charge/discharge cycling between Li_2S_x and Li_2S_y.

Considering that Li_2S_x solutions become very viscous at $x>8$ and high sulfur molarity, presumably due to association of the dissolved polysulfides [57], Figure 5.3 shows that to achieve high sulfur utilization it is essential to have the end of discharge as close to Li_2S as possible. At $x=8$, the maximum sulfur fractional utilization would be –0.87. If the end of discharge occurs, more realistically at $y=1.25$, the maximum fractional sulfur utilization becomes ~0.67. Cathode formulations that have an appreciable solubility for Li_2S_y, where $y<1.5$, are thus important development objectives.

Figure 5.3 also readily illustrates the effects of an "over-recharge." For example, a cell loaded with Li_2S_5 and initially discharged to $Li_2S_{1.43}$ would correspond to a fractional sulfur utilization of 0.50. "Over-recharging" the cell by 25%, that is, recharging the cell by 1.25×(previous discharge capacity), would then correspond to a utilization of 0.50×1.25=0.625. Such an over-recharge, although seemingly modest, with the end of discharge remaining at $Li_2S_{1.43}$ would, using Equation 5.2, bring the mean charged composition now to $x=1[0.625-(1/1.43)]=13.6$. Recharging by more than a factor of 1.25 in this example rapidly raises the S/Li ratio so that difficulties may arise. These include sulfur precipitation, formation of high-viscosity long-chain polysulfides, or initiation of irreversible side reactions due to the presence of highly oxidized polysulfide species. This, perhaps counterintuitive result, follows from the rapid lithium depletion once the lithium content of the cathode drops below that which was originally loaded.

5.7 CONDUCTIVITY OF LITHIUM POLYSULFIDE NONAQUEOUS CATHOLYTES

The rate capability of the Li/S cell will depend directly on the electrolyte conductivity of the cathode. Conductivities of nonaqueous catholytes based on aprotic solvents are significantly dependent on sulfur molarity and choice of supporting salt. Figure 5.4 shows the superiority of LiTFSI (LiN(CF$_3$SO$_2$)$_2$) salt at sulfur molarities below 6 M S for solutions of Li$_2$S$_8$ in the mixture of 1,2-dimetoxyethane (DME) and 1,3-dioxolane (Diox) [58,59] with various supporting salts. Conductivity of the catholytes with 0.5 M LiTFSI drops monotonically with increasing sulfur molarity, presumably due to significant self-association and increasing viscosity. Beyond 8 M S, the data indicate that the conductivities decline virtually independent of the added salts.

5.8 LITHIUM METAL ANODES

Cycling of the metallic lithium anode continues to be challenging because of high surface area ("mossy") lithium formation and lithium dendrite growth. Freshly formed, high surface area lithium reacts with the electrolyte, leading to irreversible loss of electrolyte and lithium metal in the cell. The application of a compressive

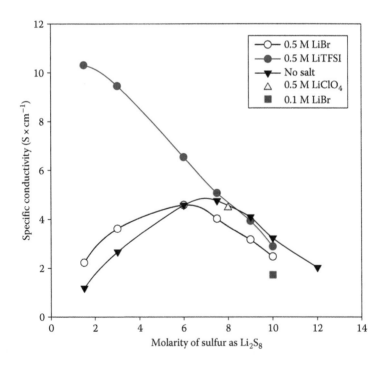

FIGURE 5.4 Ionic conductivity of polysulfide solutions in DME–Diox mixture (9:1, v/v) containing various supporting salts. (Courtesy of PolyPlus Battery Company, Berkeley, CA.)

stress on anode/cathode assembly with a microporous or other membrane separator compacts the high surface area lithium interface layer and can delay the negative consequences of dendrite shorting and moss formation, thus increasing the cycle life of the lithium anode. However, relatively high stresses, in excess of 100 kg/cm^2, may need to be exerted to have significant effect [60], posing significant battery engineering design challenges. More recent communications include the work of Crowther and West [61], who considered the effects of electrolyte composition on dendrite growth, finding that lithium dendrites form more readily with LiTFSI compared to LiPF$_6$. The advantage of LiPF$_6$ was, however, only temporary.

Dendrite formation has long been studied and modeled extensively [62], particularly for zinc electrodeposition. While various additives, such as lead or cadmium ions or surfactants, help to suppress dendrites in zinc electrodeposition [63], they have not, as yet, been reported to be of much benefit in lithium deposition, even in the absence of dissolved sulfide species. Other strategies have included the addition of LiNO$_3$ to the electrolyte [16,64], pulse charging [65,66], cesium salt addition [67], and the use of ionic liquids [68,69]. While some of these approaches have shown promise, they are not, as yet, fully successful in maintaining high lithium cycling efficiency when paired with a sulfur solvent-containing cathode in which lithium polysulfides are invariably solvated to varying degrees. These issues have also been discussed recently in a review by Scheers et al. [70].

5.9 Li/S CELLS WITH NONAQUEOUS CATHOLYTES

The polysulfide speciation distribution during charge and discharge changes not only with overall state of charge, but also with charge/discharge rates. Recent papers on the complexities of the sulfur speciation have elucidated this issue [71,72], although the effects of current densities have yet to be fully quantified *in operando*. Results indicating that the relative ratios of the products at the end of discharge are significantly current density dependent were reported by Ryu et al. [73]. Analytical speciation is significantly complicated by the fact that discharge products other than Li$_2$S do not exist as actual phases. Nevertheless, it is expedient to refer to the electrolyte compositions such as Li$_2$S$_4$ and Li$_2$S$_2$ since these overall compositions coincide with particular changes in the charge/discharge current/voltage trajectories. Typical first discharges for Li/S cells with nonaqueous catholyte are shown in Figure 5.5.

The voltage drop-off, marking the end of discharge at approximately the Li$_2$S$_2$ average catholyte composition, is associated with the precipitation of Li$_2$S and possibly other complexes [74–76]. Upon recharge, the precipitated species need to be redissolved via chemical reaction with newly formed longer-chain polysulfides. Incomplete dissolution leads to accumulation of Li$_2$S precipitate, lowering the active sulfur concentration upon repeated charge/discharge cycling. This effect is more limiting at high polysulfide molarity.

Measurements on rotating disk (RD) glassy carbon (GC) electrodes [77] have also revealed the effects of precipitation on electrode processes. Tests were performed at two different potentials, 2.25 V (vs. Li/Li$^+$), which corresponds to the solution phase region, and 1.50 V (vs. Li/Li$^+$), which corresponds to the near-end-of-discharge region where precipitation occurs. As evident from the Koutecky–Levich

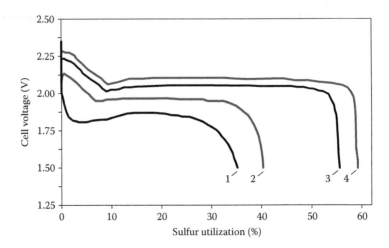

FIGURE 5.5 First discharge voltage profiles for Li/S cells with catholytes containing 7.5 M S, 0.5 M LiTFSI in DME–Diox mixture (9:1, v/v). Current densities: (1) 10 mA/cm², (2) 5.0 mA/cm², (3) 1.0 mA/cm², and (4) 0.5 mA/cm². (Courtesy of PolyPlus Battery Company, Berkeley, CA.)

plots presented in Figure 5.6a, the RDE response is classical at low overpotentials on GC electrode and describes the process controlled by diffusion of polysulfides in liquid electrolyte and an interfacial charge transfer. However, as can be seen from Figure 5.6b, at high overpotentials, corresponding to the two-phase discharge region, the response cannot be described by the same classical model. One likely interpretation would follow from the work of Weroński et al. [78], who developed a description of the RDE response at various fractional coverages of the RDE surface by solid precipitate. Higher coverage, which would be expected at lower rotation rates, lowers the current.

It may be concluded that, together with the sulfur shuttle, irreversible precipitation or slow dissolution of insoluble discharge products is a major source of cell capacity decline on cycling for the aprotic Li-S system. Clearly, a solvent system that has good solubility for "Li$_2$S$_2$" and Li$_2$S together with effective lithium protection is required. In addition, high viscosities of concentrated long-chain polysulfide solutions, possible disproportionation of long-chain polysulfides, and precipitation of sulfur on charge are potential limitations on capacity retention upon cycling.

5.10   GRAPHIC REPRESENTATION OF CHARGE–DISCHARGE PATHS

The essentials of the charge–discharge process may be readily visualized in a ternary diagram representation (Figure 5.7), similar to the isothermal ternary diagram representation used, for example, by Huggins [79]. The ternary diagram guide is presented in Figure 5.7a. The three vertices of the diagram represent pure compositions of solvent, Li metal, and sulfur. The line connecting the solvent vertex to the sulfur vertex tracks the concentration of sulfur contained in the catholyte. Similarly,

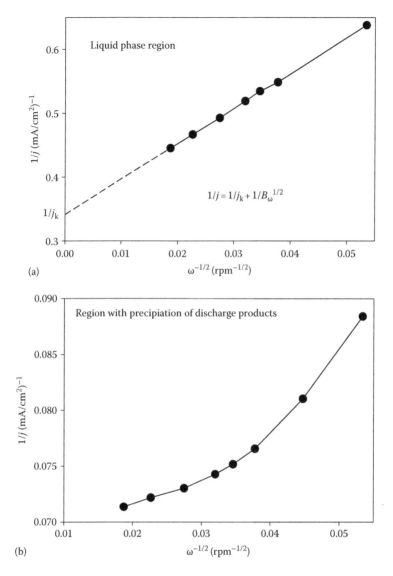

(a)

(b)

FIGURE 5.6 Koutecky–Levich plots for reduction of polysulfides on glassy-carbon RDE in electrolyte containing 3 M S as Li_2S_8, 0.5 M LiTFSI in tetraglyme at two different potentials. (a): 2.25 V vs. Li/Li$^+$; (b): 1.0 V vs. Li/Li$^+$. (Courtesy of PolyPlus Battery Company.)

the line connecting the Li vertex to the sulfur vertex tracks the Li to S ratio in the catholyte. An example of an idealized charge–discharge trajectory is shown in Figure 5.7b.

In Figure 5.7b, the solvent-Li_2S_2 line is taken to mark the solubility limit of Li_2S_{2+x}. In field A, compositions of Li_2S_x with $4 < x < 8$ are generally considered to be fully solvated. The B phase field for Li_2S_x with $2 > x \geq 4$ constitutes a more complex regime

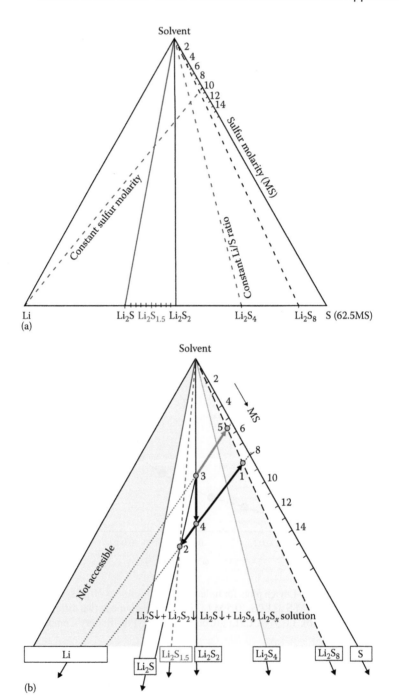

FIGURE 5.7 (a) Ternary diagram guide; (b) ternary diagram schematic illustrating quasi-equilibrium charge/discharge paths for nonaqueous Li/S cell cycling between Li_2S_8 and $Li_2S_{1.5}$ with initial 8 M S cathode. (Courtesy of PolyPlus Battery Company.)

where speciation is still under study [71,72]. The detailed speciation in B is likely to depend on the nature of the solvent, the rates of charge or discharge, and the charge–discharge history. Starting from point *1* (Li$_2$S$_8$ at 8MS) in Figure 5.7b, the arrow from point *1* to point *2* represents the global discharge path from Li$_2$S$_8$, to the average composition *2* (Li$_2$S$_{1.5}$). In this representation, the end-of-discharge composition *2*, in phase field C in Figure 5.7b, is then a three-phase pseudo-equilibrium of precipitated Li$_2$S, with a saturated solution of Li$_2$S$_2$ (point *3*) together with precipitated Li$_2$S$_2$, the relative amounts determined of which are given by the "lever rule" applied along the solvent-Li$_2$S$_2$ line and the Li$_2$S-point *3* line.

The pseudo-equilibrium recharge trajectory with full redissolution of Li$_2$S is the path *3–4–1*. The alternatively recharge path with no redissolution of Li$_2$S is the trajectory *3–5*. Continued cycling would then lead to rapid capacity fading. If Li$_2$S redissolution were slow, the recharge path would be some trajectory between these two extremes.

For partial redissolution of Li$_2$S, repeated charge-discharge cycling would produce trajectories in Figure 5.7b that move progressively closer to the solvent corner of the ternary phase diagram representation. The actual charge–discharge trajectories will depend on charge and discharge rates with path *1-(2)-3–5* corresponding to the fastest charge rates. This issue was addressed by Cheong et al. [80], relating the irreversible precipitation and location of the precipitate to the rate of charge/discharge, and observing that the precipitate tends to collect on the surface of the cathode. The rate-dependent cell capacity decline per cycle can thus be a measure of the rate of the chemical redissolution of the precipitated sulfur species. A residual "indefinite" low-capacity cycling will be determined by a dynamic balance between the rate of dissolution of the precipitated phase with the sulfur-diluted catholyte and the cell recharge current density. Experiments at our laboratories have suggested that this residual capacity is around 2 MS since at such catholyte molarity, very high cycle numbers can generally be obtained with little difficulty for aprotic solvents. It should be noted that the ternary diagram phase fields and the associated sulfur species speciation are in actuality more complex and in addition depend on polysulfide concentrations possibly due to the formation of polysulfide complexes and the dependence of the activity coefficient of the polysulfide species on their concentration. Moreover, it should be expected that the spectrum of polysulfides present depends on the charge and discharge rates. Complete characterization studies have yet to be done, but progress has been made, particularly by using *in operando* x-ray absorption spectroscopy [81,82]. The relative simplicity of the ternary representation can help, however, in providing a visual summary of some of the complications of the Li/S cell operation.

5.11 Li/S CELLS WITH AQUEOUS CATHOLYTES

As is evident from Figure 5.3, obtaining high cathode energy densities requires the ability to reversibly discharge the Li/S cell as close to Li$_2$S as possible. This is very difficult to achieve using catholytes based on aprotic solvents since no aprotic solvents are known to dissolve Li$_2$S to any appreciable degree. The approach of Visco et al. [83] has been to explore aqueous catholytes. A similar system, but with low sulfur molarity, was subsequently reported by Zhou et al. [84]. In aqueous media, Li_2S will hydrolyze, forming LiSH and LiOH, leading to solutions with high pH where the formation of H_2S is suppressed. Aqueous polysulfide electrolytes have also been considered by Licht and Peramunage for the aluminum/sulfur system [85]. The hydrolysis of Li_2S can be written as $Li_2S + H_2O \rightarrow LiHS + LiOH$ [86].

Formation of H_2S is greatly suppressed at pH > 9, and the effective solubility (via hydrolysis) of the Li_2S is likely limited to about 3M by the solubility limit of hydrated LiOH. At the end of a full discharge, the presence of dissolved LiSH and LiOH, precipitated LiOH monohydrate, and possibly, precipitated Li_2S, should be expected.

The voltage window of aqueous Li/S cells during cell discharge is limited by the process of electrochemical hydrogen evolution from water. The use of carbon matrices and cathode current collectors with large hydrogen overpotentials [87] allows for deeper discharge of the Li/S aqueous cells. The cell charging voltage is limited by possible electrochemical formation of oxysulfur species, in particular thiosulfates. Although the composition of the aqueous catholyte near the end of a deep discharge (Li_2S_y, where $y < 2$) appears to be a complex mix, the key advantage of this approach is that in aqueous systems the concentrations of active sulfur species can be significantly higher than those in their nonaqueous counterparts. In the Li/S aqueous cells, both the significant concentrations of active sulfur species present at the end of discharge and fast chemical dissolution of discharge precipitate in solution are beneficial for fast charging, even after the deep discharge. As seen in Figure 5.8, the voltage profiles are notably different from those for the nonaqueous Li/S cells.

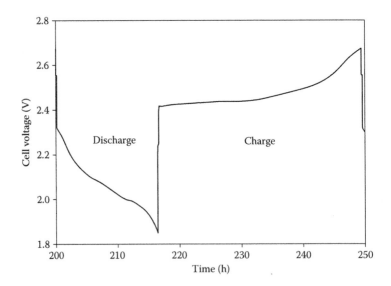

FIGURE 5.8 Discharge/charge voltage profiles for Li/S cell with water-based catholyte and protected Li electrode (5th cycle). Cell was assembled in the charged state with 12 M S as Li_2S_5 (electrolyte-to-sulfur ratio is 2.6:1 mL/g). Discharge current density is 1.0 mA/cm^2, and charge current density is 0.5 mA/cm^2. Discharge capacity is 16.4 mAh/cm^2 and sulfur utilization is 57%. (Courtesy of PolyPlus Battery Company.)

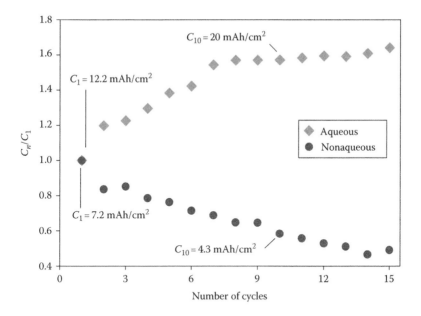

FIGURE 5.9 Comparison of capacity evolution on cycling for Li/S cells with aqueous and nonaqueous catholytes. Discharge current density is 1.0 mA/cm², and charge current density is 0.5 mA/cm². Aqueous catholyte contains 12 M S as Li_2S_5, and nonaqueous catholyte contains 10 M S as Li_2S_8, 0.5 M LiTFSI in DME-Diox. (Courtesy of PolyPlus Battery Company.)

In particular, the discharge curve does not have a pronounced plateau corresponding to generation of Li_2S_y species with $y \leq 4$ on discharge. The sharp voltage downturn that marks the end of discharge would suggest that, similar to the nonaqueous Li/S system, precipitation of discharge products determines the end of the discharge. Cycling of aqueous Li/S cells at high sulfur concentrations has shown encouraging results, in that the capacity decline is significantly reduced. Figure 5.9 shows a comparison of capacity evolution on cycling for aqueous and nonaqueous Li/S cells.

5.12 CONCLUSION

Lithium metal battery systems continue to show considerable progress and promise. A number of challenges remain before we can expect to see these systems compete in commercial applications that are currently employing lithium-ion systems with solid oxide cathodes. Among these challenges is the development of high cycle life lithium metal anodes with low lithium excess. Solvents are necessary for the operation of the sulfur cathodes, which in turn necessitates a fully effective anode protection. Polymeric membranes or chemically developed SEI layers have, to date, not been fully effective; in contrast, inorganic glass, glass–ceramic, or ceramic membranes have been most effective and have demonstrated the elimination of the sulfur redox shuttle. High sulfur utilization at the elevated sulfur molarities necessary for achieving high energy density is limited by the solubility of the polysulfides on charge and precipitation of discharge products. Enhanced solubility may be reached for different solvent choices, and among these, aqueous polysulfide cathodes have shown considerable promise.

Much still needs to be done on clarifying the sulfur cathode chemistry under dynamic conditions. Extensive efforts on speciation as a function of the dynamic parameters could be of considerable help in reaching high sulfur utilization at practical rates.

ACKNOWLEDGMENTS

This work was funded in part by ARPA-E of the U.S. Department of Energy. The authors also acknowledge helpful discussions with the PolyPlus staff, especially Hannah Ray, Vitaliy Nimon, Kirill Pridatko, Nikolay Goncharenko, Bruce Katz, and May-Ying Chu.

REFERENCES

1. Herbert, D. and J. Ulam. 1958. Electric dry cells and storage batteries. US Patent 3,043,896, filed November 24, 1958, and issued July 10, 1962.
2. Compiled by PolyPlus Battery Company, Fall 2014, from https://www.scholar.google.com. Search terms: Lithium Sulfur Battery and Li-S Battery.
3. Rauh, R. D., K. M. Abraham, G. F. Pearson, J. K. Surprenant, and S. B. Brummer. 1979. A lithium/dissolved sulfur battery with an organic electrolyte. *J Electrochem Soc* 126:523–527.
4. Song, M.-K., Y. Zhang, and E. J. Cairns. 2013. A long-life, high-rate lithium/sulfur cell: A multifaceted approach to enhancing cell performance. *Nano Lett* 13:5891–5899.
5. Dey, A. N. and P. Bro. 1968. Light metal-sulfur organic electrolyte cell. US Patent 3,806,369, filed November 13, 1968, and issued April 23, 1974.
6. Visco, S. J., M.-Y. Chu, and L. C. De Jonghe. 1999. Overcharge protection systems for rechargeable batteries. US Patent 5,882,812, filed January 14, 1997, and issued March 16, 1999, and related patents.
7. Mikhaylik, Y. and J. Akridge. 2004. Polysulfide shuttle study in the Li/S battery system. *J Electrochem Soc* 151:A1969–A1976.
8. Yamin, H. and E. Peled. 1983. Electrochemistry of a nonaqueous lithium/sulfur cell. *J Power Sources* 9:281–287.
9. Ji, X., K. T. Lee, and L. F. Nazar. 2009. A highly ordered nanostructured carbon–sulphur cathode for lithium–sulphur batteries. *Nat Mater* 8:500–506.
10. Evers, S. and L. Nazar. 2012. Graphene-enveloped sulfur in a one-pot reaction: A cathode with good coulombic efficiency and high practical sulfur content. *Chem Commun* 48:1233–1235.
11. Seh, Z., W. Li, J. Cha, G. Zheng, Y. Yang, M.T. McDowell, P.-C. Hsu, and Y. Cui. 2013. Sulphur-TiO$_2$ yolk-shell nanoarchitecture with internal void space for long-cycle lithium-sulfur batteries. *Nat Commun* 4:1331.
12. Skotheim, T. A., B. Trofimov, A. G. Malkina, and I. P. Koralev. 1995. Electroactive high storage capacity polyacetylene-co-polysulfur materials and electrolytic cells containing same. US Patent 5,529,860, filed June 7, 1995, and issued June 25, 1996.
13. Hagen, M., P. Fanz, and J. Tübke. 2014. Cell energy density and electrolyte/sulfur ratio in Li-S cells. *J Power Sources* 264:30–34.

14. Zhao, Y., Y. Zhang, D. Gosselink, T. N. L. Doan, M. Sadhu, H.-J. Cheang, and P. Chen. 2012. Polymer electrolytes for lithium/sulfur batteries. *Membranes* 2:553–564.

15. Zhang, S. S. 2012. Improving the cyclability of liquid electrolyte Lithium/sulfur batteries by optimizing electrolyte/sulfur ratio. *Energies* 5:5190–5197.

16. Mikhaylik, Y. V. 2004. Electrolytes for lithium sulfur cells. US Patent 7,354,680 B2, filed January 6, 2004, and issued April 8, 2008.

17. Zhang, S. S. and J. A. Read. 2012. A new direction for the performance improvement of rechargeable lithium/sulfur batteries. *J Power Sources* 200:77–82.

18. Bates, J. B., N. J. Dudney, G. R. Gruzalski, R. A. Zuhr, A. Choudhury, and C.F. Luck. 1992. Electrical properties of amorphous lithium electrolyte thin films. *Solid State Ionics* 53–56:647–654.

19. Aono, H., E. Sugimoto, Y. Sadaoka, N. Imanaka, and G. Adachi. 1989. Ionic conductivity of lithium titanium phosphate ($Li_{1+x}M_xTi_{2-x}(PO4)_3$, M=Al, Sc, Y, and La) systems. *J Electrochem Soc* 136:590–591.

20. Fu, J. 1997. Superionic conductivity of glass-ceramics in the system Al_2O_3-TiO_2-P_2O_5. *Solid State Ionics* 96:195–200.

21. Skotheim, T. A., C. J. Sheehan, Y.V. Mikhaylik, and J. Affinito. 2007. Lithium anodes for electrochemical cells. US Patent 8,753,771, filed July 23, 2007, and issued June 17, 2014.

22. Murugan, R., V. Thangadurai, and W. Weppner, 2007. Fast lithium ion conduction in garnet-type $Li_7La_3Zr_2O_{12}$. *Angew Chem Int Ed* 46:7778–7781.

23. Ma, C., E.M. Rangasamy, Ch. Liang, J. Sakamoto, K. L. More, and M. Chi. 2015. Excellent stability of a lithium-ion-conducting solid electrolyte upon reversible Li^+/H^+ exchange in aqueous solutions. *Angew Chem Int Ed* 54:129–133.

24. Visco, S. J., B. D. Katz, Y. S. Nimon, and L. C. De Jonghe. 2004. Protected active metal electrode and battery cell structures with non-aqueous interlayer architecture. US Patent 7,282,295, filed April 14, 2004, and issued October 16, 2007; Visco, S. J. and Y. S. Nimon. 2004. Active metal/aqueous electrochemical cells and systems. US Patent 7,645,543 B2, filed February 3, 2004, and issued January 12, 2010; Visco, S. J., B. D. Katz, Y. S. Nimon, and L. C. De Jonghe. 2010. Protected active metal electrode and battery cell with ionically conductive protective architecture. US Patent 8,293,398 B2, filed September 22, 2010, and issued October 23, 2012.

25. Wang, X., Y. Hou, Y. Zhu., Y. Wu, and R. Holze. 2013. An aqueous rechargeable lithium battery using coated Li metal as anode. *Sci Rep* 3:1401.

26. Zhou, H., Y. Wang, H. Li, and P. He. 2010. The development of a new type of rechargeable batteries based on hybrid electrolytes. *ChemSusChem* 3:1009–1019.

27. Kummer, J. T. and N. Weber. 1966. Battery having molten alkali metal anode and molten sulfur cathode. 1968. US Patent 3,413,150, filed September 28, 1966, and issued November 26, 1968.

28. Bragg, W. L., C. Gottfried, and J. West. 1931. The structure of beta alumina. *Z Kristallogr* 77:255–274.

29. Blass, H., P. Charleroi, and H. Baumann. 1934. Furnace lining and material therefor. US Patent 2,043,029, filed April 19, 1934, and issued June 2, 1936.

30. Yao, Y.-F. Y. and J. Kummer. 1967. Ion exchange properties of and rates of ionic diffusion in beta-alumina. *J Inorg Nucl Chem* 29:2453–2466; Kummer, J. 1972. β-alumina electrolytes. *Prog Solid State Chem* 7:141–175.

31. Wen, Z., Y. Hu, X. Wu, J. Han, and Z. Gu. 2013. Main challenges for high performance NAS battery: Materials and interfaces. *Adv Funct Mater* 23:1005–1018.

32. Lu, X., G. Coffey, K. Meinhardt, V. Sprenkle, Z. Yang, and J. P. Lemmon. 2010. High power planar sodium-nickel chloride battery. *ECS Trans* 28:7–13.

33. Coors, W. G., J. H. Gordon, and S. G. Menzer. 2010. Electrochemical cell comprising ionically conductive membrane and porous multiphase electrode. US patent application US20100297537 A1, filed February 3, 2010.

34. Pearson, G. and P. L. Robinson. 1930. The polysulfides of the alkali metals. Part I. *J Chem Soc* 133:1473–1497.

35. Weber, N. and J. T. Kummer. 1967. Sodium-sulfur secondary batteries. *Proc 21st Annu Power Sources Conf* 21:37–39.

36. Sulfur utilization is defined as the ratio of discharge capacity to total sulfur capacity (which is the sum of all forms of sulfur capacity in the cathode including capacity due to elemental sulfur and polysulfides).

37. Okamoto H. 1995. The Li-S (lithium-sulfur) system. *J Phase Equilib* 16:94–97.

38. Brennan, M. P. J. 1990. The failure of beta-alumina electrolyte by a dendritic penetration mechanism. *Electrochim Acta* 25:621–627.

39. Monroe, C. and J. Newman. 2003. Dendrite growth in lithium/polymer systems. *J Electrochem Soc* 150:A1377–A1384.

40. Monroe, C. and J. Newman. 2005a. The effect of interfacial deformation on electrodeposition kinetics. *J Electrochem Soc* 150:A396–A404.

41. Monroe, C. and J. Newman. 2005b. The impact of elastic deformation on deposition kinetics at lithium-polymer interfaces. *J Electrochem Soc* 151:A880–A886.

42. Stark, J. K., Y. Ding, and P. A. Kohl. 2013. Nucleation of electrodeposited lithium metal: Dendritic growth and the effect of co-deposited sodium. *J Electrochem Soc* 160:D337–D342.

43. De Jonghe, L. C., L. Feldman, and A. Buechele. 1981. Failure modes of Na-beta alumina. *Solid State Ionics* 5:267–269.

44. Sargent, P. and M. Ashby. 1984. Deformation maps for alkali metals. *Scr Metall* 18:145–150.

45. Ishiguro, K., Y. Nakata, M. Matsui, I. Uechi, Y. Takeda, O. Yamamoto, and N. Imanishi. 2013. Stability of Nb-doped cubic $Li_7La_3Zr_2O_{12}$ with lithium metal. *J Electrochem Soc* 160:A1690–A1693; Ishiguro, K., H. Nemori, S. Sunahiro Y. Nakata, R. Sudo, M. Matsui, Y. Takeda, O. Yamamoto, and N. Imanishi. 2014. Ta-doped $Li_7La_3Zr_2O_{12}$ for water-stable lithium electrode of lithium-air batteries. *J Electrochem Soc* 161:A668–A674.

46. Binner, J. and R. Stevens. 1985. The improvement of the mechanical properties of polycrystalline beta-alumina *via* the use of zirconia particles containing stabilizing oxide additions. *J Mater Sci* 20:3119–3124.

47. May, G. 1978. The development of beta-alumina for use in electrochemical cells: A survey. *J Power Sources* 3:1–22.

48. Hitchcock, D. and L. C. De Jonghe. 1983. Fracture toughness anisotropy of sodium β-alumina. *Commun Am Ceram Soc* 66:c204–c205.

49. Mizuno, F., F. Hayashi, A. Tadanaga, and K. Tatsumisago. 2005. New, highly ion-conductive crystals precipitated from Li_2S-P_2S_5 glasses. *Adv Mater* 17:918–921.

50. Kamaya, N., K. Homma, Y. Yamakawa, M. Hirayama, R. Kanno, M. Yonemura, T. Kamiyama et al. 2011. A lithium superionic conductor. *Nat Mater* 10:682–686.

51. Kuhn, A., O. Gerbig, C. Zhu, F. Falkenberg, J. Maier, and B. V. Lotsch. 2014. Ultrafast Li electrolytes based on abundant elements: $Li_{10}SnP_2S_{12}$ and $Li_{11}Si_2PS_{12}$, arxiv.org/pdf/1402.4586. (accessed Fall 2014).

52. Nagao, M., A. Hayashi, M. Tatsumisago, T. Kanetsuku, T. Tsuda, and S. Kuwabata. 2013. In situ SEM study of a lithium deposition and dissolution mechanism in a bulk-type solid-state cell with a Li_2S–P_2S_5 solid electrolyte. *Phys Chem Chem Phys* 15:18600–18606.

53. Bates, J. B., N. J. Dudney, G. R. Gruzalski, R. A. Zuhr, A. Choudhury, and C. F. Luck. 1993. Fabrication and characterization of amorphous lithium electrolyte thin films and rechargeable thin-film batteries. *J Power Sources* 43–44:103–110.

54. Dudney, N. J. 2008. Thin film micro-batteries. *Interface* 17:44–48.

55. Bobeth M., M. Gutkin, W. Pompe, and A. E. Romanov. 1998. Modeling of vacancy diffusion and pore formation during parabolic oxide growth. *Phys Stat Sol (a)* 165: 165–184.

56. Zu, C., Y. Fu, and A. Manthiram. 2013. Highly reversible Li/dissolved polysulfide batteries with binder-free carbon nanofiber electrodes. *J Mater Chem A* 1:10362–10367.

57. Kolosnitsyn, V. S., E. V. Kuzmina, and E. V. Karaseva. 2009. Effect of lithium salts on the physicochemical properties of lithium polysulphide solutions in sulfolane. Abstract #184, *215th Electrochemical Society Meet*, San Francisco, CA, May 24–29, 2009.

58. Nimon, Y., S. J. Visco, and M.Y. Chu. 1999. Dioxolane as a protector for lithium electrodes. US Patent 6,225,002 B1, February 5, 1999, and issued May 1, 2001.

59. Kim, S., Y. Jung, and H. S. Lim. 2004. The effect of solvent component on the discharge performance of lithium–sulfur cell containing various organic electrolytes, *Electrochim Acta* 50:889–892.

60. Hirai, T., I. Yoshimatsu, and J.-i. Yamaki. 1994. Influence of electrolyte on lithium cycling efficiency with pressurized electrode stack. *J Electrochem Soc* 141:611–614.

61. Crowther, O. and A. C. West. 2008. Effect of electrolyte composition on lithium dendrite growth. *J Electrochem Soc* 155:A806–A811.

62. Barton, J. L. and J. O'M. Bockris. 1962. The electrolytic growth of dendrites from ionic solutions. *Proc R Soc A* 268:485–505.

63. Kan, J., H. Xue, and S. Mu. 1998. Effect of inhibitors on Zn-dendrite formation for zinc-polyaniline secondary battery. *J Power Sources* 74:113–116.

64. Zhang, S. S. 2012. Role of $LiNO_3$ in rechargeable lithium/sulfur battery. *Electrochim Acta* 70:344–348.

65. Mayers, M. Z., J. W. Kaminski, and T. F. Miller, III. 2012. Suppression of dendrite formation via pulse charging in rechargeable lithium metal batteries. *J Phys Chem C* 116:26214–26221.

66. Park, M. S., S. B. Ma, D. J. Lee, D. Im, S.-G. Doo, and O. Yamamoto. 2014. A highly reversible lithium metal anode. *Sci Rep* 4:3815.

67. Ding, F., W. Xu, G. L. Graff, J. Zhang, M. L. Sushko, X. Chen, Y. Shao et al. 2013. Dendrite-free lithium deposition via self-healing electrostatic shield mechanism. *J Am Chem Soc* 135:4450–4456.

68. Yoon, H., P. C. Howlett, A. S. Best, M. Forsyth, and D. R. MacFarlane. 2013. Fast charge/discharge of Li metal batteries using an ionic liquid electrolyte. *J Electrochem Soc* 160: A1629–A1637.

69. Blomberg, E. 2012. Redox behavior of Li-S cell with PP14-TFSI ionic liquid electrolyte. Master thesis, Chalmers University of Technology, http://publications.lib.chalmers.se/records/fulltext/160632.pdf. (accessed Fall 2014).

70. Scheers, J., S. Fantini, and P. Johansson. 2014. A review of electrolytes for lithium-sulphur batteries. *J Power Sources* 255:204–218.

71. Barchasz, C., F. Molton, C. Duboc, J.-C. Lepretre, S. Patoux, and F. Alloin. 2012. Lithium/sulfur cell discharge mechanism: An original approach for intermediate species identification. *Anal Chem* 84:3973–3980.

72. Lowe, M., J. Gao, and H. Abrũna. 2014. Mechanistic insights into operational lithium–sulfur batteries by in situ X-ray diffraction and absorption spectroscopy. *RSC Adv* 4:18347.

73. Ryu, H. S., Z. Guoa, H. J. Ahn, G. B. Cho, and H. Liua. 2009. Investigation of discharge reaction mechanism of lithium/liquid electrolyte/ sulfur battery. *J Power Sources*. 189:1179–1183.

74. Barchasz, C., J.-C. Leprêtre, F. Alloin, and S. Patoux. 2012. New insights into the limiting parameters of the Li/S rechargeable cell. *J Power Sources* 199:322–330.

75. Kumaresan, K., Y. Mikhaylik, and R. E. White. A mathematical model for a lithium–sulfur cell. *J Electrochem Soc* 155:A576–A582.

76. Fronczek, D. and W. G. Bessler. 2013. Insight into lithium–sulfur batteries: Elementary kinetic modeling and impedance simulation. *J Power Sources* 244:183–188.

77. Visco, S., E. Nimon, B. Katz, L. C. De Jonghe, and M.-Y. Chu. 2003. Presented at the *First International Conference on Polymer Batteries and Fuel Cells* (PBFC-1), Jeju Island, Korea, June 1–6, 2003.

78. Weroński, P., M. Nosek, and P. Batys. 2013. Limiting diffusion current at rotating disk electrode with dense particle layer. *J Chem Phys* 139:124705–124709.

79. Huggins, R. A. 2009. Ternary electrodes under equilibrium or near-equilibrium conditions. In *Advanced Batteries: Materials Science Aspects*. Springer: New York.

80. Cheong, S.-F., K.-S. Ko, S.-W. Kim, E.-y. Chin, and H.-T. Kim. 2003. Rechargeable lithium sulfur battery. II. Rate capability and cycle characteristics. *J Electrochem Soc* 150: A800–A805.

81. Cuisinier, M., P.-E. Cabelguen, S. Evers, G. He, M. Kolbeck, A. Garsuch, T. Bolin, M. Balasubramanian, and L. F. Nazar. 2013. Sulfur speciation in Li-S batteries determined by operando X-ray absorption spectroscopy. *J Phys Chem Lett* 4:3227–3232.

82. Wujcik, K. H., J. Velasco-Velez, C. H. Wu, T. Pascal, A. A. Teran, M. A. Marcus, J. Cabana et al. 2014. Fingerprinting lithium-sulfur battery reaction products by X-ray absorption spectroscopy. *J Electrochem Soc* 161:A1100–A1116.

83. Visco, S., Y. Nimon, B. Katz, L. C. De Jonghe, N. Goncharenko, and V. Loginova. 2012. Aqueous electrolyte lithium sulfur batteries. US Patent 8,828,575, filed November 14, 2012, and issued September 9, 2014; Visco, S. J., Y. S. Nimon, B. D. Katz, L. C. De Jonghe, N. Goncharenko, and V. Loginova. 2012. Electrolyte compositions for aqueous electrolyte lithium sulfur batteries. US Patent 8,828,574, filed May 18, 2012 and issued September 9, 2014.

84. Li, N., Z. Weng, Y. Wang, F. Li, H.-M. Cheng, and H. Zhou. 2014. An aqueous dissolved polysulfide cathode for lithium–sulfur batteries. *Energy Environ Sci* 7:3307–3312.

85. Licht, S. and D. Peramunage. 1993. Novel aqueous aluminum/sulfur batteries. *J Electrochem Soc* 140:L4–L6.

86. Licht, S. and J. Stuart. 1997. Disproportionation of aqueous sulfur and sulfide: Kinetics of polysulfide decomposition. *J Phys Chem B* 101:2540–2545.

87. Visco, S. J., Y. S. Nimon, B. D. Katz, L. C. De Jonghe, N. Goncharenko, and V. Loginova. 2012. Electrode structures for aqueous electrolyte lithium sulfur batteries. US Patent 8,828,573, filed April 5, 2012 and issued September 9, 2014.

6 Vanadium–Air Redox Flow Batteries
Fundamentals and Applications

Vladimir Neburchilov and Jiujun Zhang

CONTENTS

6.1 INTRODUCTION

Vanadium–air redox flow batteries (VARFBs) (Figure 6.1a) are modified vanadium–redox flow batteries (VRFBs) [1,2] (Figure 6.1b) that demonstrate higher energy density, 20% higher standard cell voltage [1–10] after replacement of the positive electrode redox couple VO_2^+/VO^{2+} with an oxygen reduction reaction (ORR), and reduced cost due to elimination of the most expensive part of the V^{5+}/V^{4+} electrolyte (43% of the total cost of a VARFB, as shown in Figure 6.2 [9,11]). A bidirectional VARFB system consists of one tank with a circulated anolyte V^{2+}/V^{3+} solution, a positive air-cathode, an inert negative electrode, an ion exchange membrane (IEM) separating the electrodes and providing proton transport, a sulfuric acid catholyte, inert positive electrodes, a current collector, bipolar plates, frames, heat exchangers, and a power conversion system [8].

In general, both VRFBs and VARFBs can be used for stationary energy storage systems, carrying out peak leveling, energy shifting, grid flexibility and improved distribution efficiency, and so on. However, they have the common disadvantage of low energy density (20–30 Wh/kg) when compared with other type of batteries due

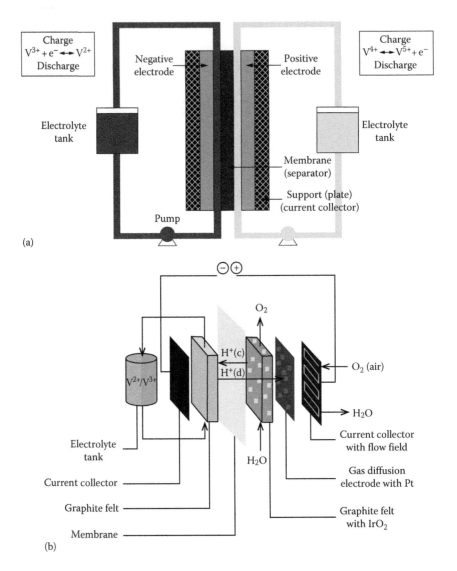

FIGURE 6.1 (a) Scheme of all-vanadium redox flow battery (VRFB) and (b) vanadium-air redox flow battery (VARFB). (Note that VARFB does not need a catholyte tank, because the O_2 comes from air and the anolyte reactants (H_2SO_4 and VSO_4) come from the anolyte tank). ([a]: Reprinted from *J. Electrochim. Acta*, 47, Fabjan, Ch., Garche, J., Harrer, B., Jörissen, L., Kolbeck, C., Philippi, F., Tomazic, G., and Wagner, F., The vanadium redox-battery: An efficient storage unit for photovoltaic systems, 825–831. Copyright 2011, with permission from Elsevier; [b]: Reprinted from *J. Power Sources*, 273, Austing, G., Kitcher, C.N., Hammer, E.-M., Komsiyska, L., and Wittstock, G., Study of an unitized bidirectional vanadium/air redox flow battery comprising a two-layered cathode, 1163–1170. Copyright 2015, with permission from Elsevier.)

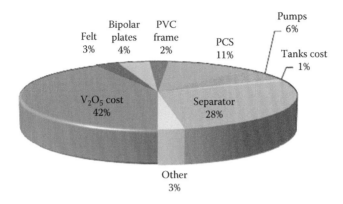

FIGURE 6.2 Cost breakdown of VRFB by components. (Reprinted from *J. Power Sources*, 24, Viswanathan, V., Crawford, A., Stephenson, D., Kim, S., Wang, W., Li, B., Coffey, G. Ed., Thomsen, G.G., Balducci, P., Kintner-Meyer, M., and Sprenkle, V., Cost and performance model for redox flow batteries, 1040–1051. Copyright 2014, with permission from Elsevier.)

to the limited solubility of vanadium species, relatively expensive vanadium electrolyte, and vanadium crossover through the membrane in the cell. Although their energy density could be heightened by increasing the vanadium solubility using special additives, optimizing the cell design with the use of zero-gap cell design (with low cell ohmic resistance), the increased energy density is still limited. Compared to VRFBs, VARFBs have several advantages, as follows [1]:

- Both higher energy density (which can be as high as 60 Wh/kg) and standard cell voltage than those (<30 Wh/kg) of VRFBs due to the significantly reduced weight and volume of the cathodic container
- The absence of expensive VRFB VO_2^+/VO^{2+} redox material in vanadium
- An unlimited source of oxygen (ambient air) for the discharging reaction

Regarding the cell structure, the components of VARFBs' anode, anolyte (sulfuric acid), and membrane are typically the same as those used in VRFBs, but the cathode (bifunctional electrode [BE]) and cathodic reaction (ORR) are different. The cathode is the BE (gas diffusion air electrode) for both the ORR and oxygen evolution reaction (OER). In this way, the catholyte tank in VARFB is not needed, which contributes to the increase in energy density.

The main technical shortcomings of the state-of-the-art VARFBs are their maturity, including low durability and the absence of cost-effective BE s, as well as unoptimized cell architectures with low ohmic (contact) resistance. For example, the advanced rechargeable VARFB (EWE Research Centre, Germany, 2015) had only a maximum energy efficiency of 41.7% at the current density of 20 mA/cm², which seemed insufficient for grid-scale energy storage application, probably due to their nonoptimized BE architecture and cell design (high vanadium crossover and contact resistance).

6.2 ELECTROCHEMISTRY OF VARFB

Regarding the electrochemistry of VARFB, the electrode reactions can be described by the following reactions:

Positive electrode reaction:

$$4VSO_4 + 2H_2SO_4 \leftrightarrow 2(V)_2(SO_4)_3 + 4e^- + 4H^+ \quad E_n^o = -0.255 \text{ V(SHE)} \quad (6.1)$$

Negative electrode reaction:

$$O_2 + 4H^+ + 4e^- \leftrightarrow 2H_2O \quad E_p^o = 1.229 \text{ V(SHE)} \tag{6.2}$$

The overall cell reaction:

$$4VSO_4 + O_2 + 2H_2SO_4 \leftrightarrow 2(V)_2(SO_4)_3 \quad E_{cell}^o = 1.484 \text{ V(SHE)} \tag{6.3}$$

where E_n^o, E_p^o, and E_{cell}^o are the thermodynamic positive electrode potential, negative electrode potential, and the cell voltage at standard conditions, respectively. Note that the electrode potentials/cell voltage of these reactions are thermodynamically expected and do not necessarily equal the practical values due to the slower electrode reaction kinetics. Under a current load of the cell, both the positive and negative electrodes will have overpotentials, which make the cell voltage lower than that expected by Reaction 6.3.

Regarding the VARFBs, the electrochemical reaction of the negative electrode, made typically of activated carbon or graphite, is the same as that in VRFBs and probably has similar reaction kinetics. The kinetic parameters for the negative reaction (6.1) in VARFBs are given in Table 6.1 [1].

At the current state of technology, VARFBs' performance is lower than that of other metal–air batteries due to several factors, such as higher reaction overpotentials,

---

TABLE 6.1

Kinetic Parameters of Redox Couple $VSO_4/(V)_2(SO_4)_3 (V^{2+}/V^{3+})$ in VARFB

Redox Couple	α	k_o (cm/s)	Electrode
V^{3+}/V^{2+}	~0.5	4×10^{-3}	Hg
	~0.5	$\gg 10^{-5}$	Graphite

Source: Reprinted from *Electrochim. Acta,* 101, Parasuramn, A., Lim, T.M., Menictas, C., and Skyllos-Kazacos, M., Review of material research and development for vanadium redox flow battery applications, pp. 27–40. Copyright 2013, with permission from Elsevier.

ohmic voltage losses from battery components, a side reaction (hydrogen evolution reaction [HER] at negative electrode), and vanadium crossover. However, VARFBs do not undergo a visible HER (during the charging of the cell up to a 90% state of charge [SOC]), indicating a high charging efficiency.

Regarding the negative electrode materials of VARFBs, carbon-based materials are well suited because they can offer a wide operation potential window, chemical stability, and a reasonable cost. However, the electrode kinetics need to be further improved for practical applications. For example, surface treatment to reduce the overpotential of the negative electrode for the V^{2+}/V^{3+} redox reaction is necessary. For improving VARFBs' energy density, reducing the electrode mass and the volume of the electrolyte container by increasing the reactant's concentration seems to be effective.

Regarding the positive electrode of VARFBs, the positive electrode reaction (ORR) has a 20% higher standard potential than that of VRFBs (V^{4+}/V^{5+} redox reaction). To facilitate both the ORR (discharge process) and OER (charge process), the positive electrode can be designed to contain some mixed catalysts or alloy (such as Pt-Ir-based catalyst) to form the bifunctional catalyst layer, in which one catalyst component is functioning for OER, and the other for ORR [10]. In fact, the most critical component of VARFBs is this bifunctional positive electrode with bifunctional catalysts. The majority of efforts in VARFB research and development are focused on this electrode and its catalysts at the current state of technology. BEs have several designs, such as a single electrode with a single catalytic layer, separate electrodes with a single catalytic layer, and a single electrode with multiple catalytic layers.

6.3 DESIGN OF VARFBs

Several VARFB designs have been developed during the last 20 years [2–8] since their first invention in 1994 by Kaneko et al. [3] in the effort to increase both the energy density with a high concentration of vanadium electrolyte and the stability of their performance [1]. Menitas and Skyllas-Kazacos [4] developed a non rechargeable five-cell stack of vanadium–oxygen fuel cell (VOFC) that could discharge for 120 h. In their design, Nafion® 112/117 membranes were used to construct the membrane electrode assembly (MEA). The main challenge of this kind of electrode design may be the delamination of the catalyst layer from the catalyst-coated membrane (CCM), limiting the lifetime of battery. Hosseiny et al. [4–6] developed a modular design for a VARFB called the VOFC, with two membrane electrode assemblies (2MEA design) for charging (with an ORR catalyst layer containing a Pt/C catalyst) and discharging (with an OER catalyst layer containing a IrO_2/TiO_2 catalyst). However, this VOFC seemed to overcome the issue of insufficient energy efficiency of 26.7% at $T = 40°C$ in 3 M H_2SO_4 by eliminating the side hydrogen evolution during the discharge process. Noack et al. [8,9] tried to improve the performance of such VOFCs using a cell with a geometric active area of 280 cm^2, consisting of two half-cells separated by two membranes (Nafion® 115 and 211) and an intermediate chamber (gap flow frame) containing circulated 2 M H_2SO_4 solution (Figure 6.3).

2 M sulfuric acid solution was gassed externally with air to maintain a redox potential of the V^{2+} electrolyte above +0.1 V (NHE) and avoid the side HER [8,9].

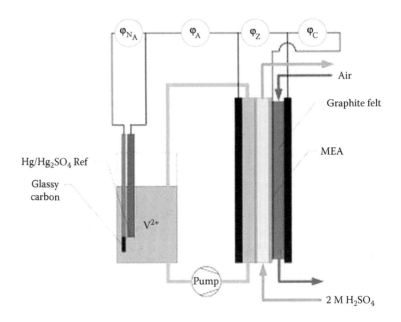

FIGURE 6.3 Shematic setup of the VOFC (50 cm²). (Reprinted from *J. Power Sources*, 259, Noack, J., Cremers, C., Bayer, D., Tübke, J., and Pinkwart, K., A coupled-physics model for the vanadium oxygen fuel cell, 125–137. Copyright 2014, with permission from Elsevier.)

The VOFC has an anodic half-cell and was constructed by a PVC flow frame graphite foil and several layers of graphite felts, and this half-cell was separated by a Nafion 115® membrane. The cathodic half-cell consists of a flow-through frame, graphite foil, graphite felt, gas diffusion layer (GDL), and Pt CCM. Anolyte $V^{2+/3+}$ was prepared by electrolyzing the solution containing 0.8 M $VOSO_4$, 0.4 M $V_2(SO_4)_3$, 2 M H_2SO_4, and 0.05 H_3PO_4 until the redox potential of 1.1 V (Hg/Hg_2SO_4). The key element of this VOFC/2 M H_2SO_4 design was the intermediate chamber formed by a gap flow frame located in the graphite cavity between two half-cells, which was separated by two membranes with the purpose of minimizing the diffusion of V^{2+} (crossover) from the anodic half-cell onto the Pt catalyst in CCM of cathodic cell for reducing the VOFC performance degradation. Thermodynamically, the proton could oxidize V^{2+} into V^{3+} to produce hydrogen, as shown in Reaction 6.4:

$$2VSO_4 + H_2SO_4 \leftrightarrow (V)_2(SO_4)_3 + H_2 \tag{6.4}$$

It is interesting to note that the side Reaction 6.4 on a Pt catalyst of the negative electrode can indeed reduce cell efficiency. To reduce this side Reaction 6.4, using air to blow the anolyte showed a significant effect on the discharging potential. In addition, the use of two membranes in the cell could suppress the crossover of V^{2+} ions to the Pt/C cathode. VOFC/2 M H_2SO_4 with 280 and 50 cm² active surface areas, respectively, could demonstrate an improved voltage efficiency [9].

Figure 6.4 shows the performance of a VOFC/2 M H_2SO_4 cell at different air (a) and electrolyte (b) flow rates, respectively. It can be seen that increasing both air

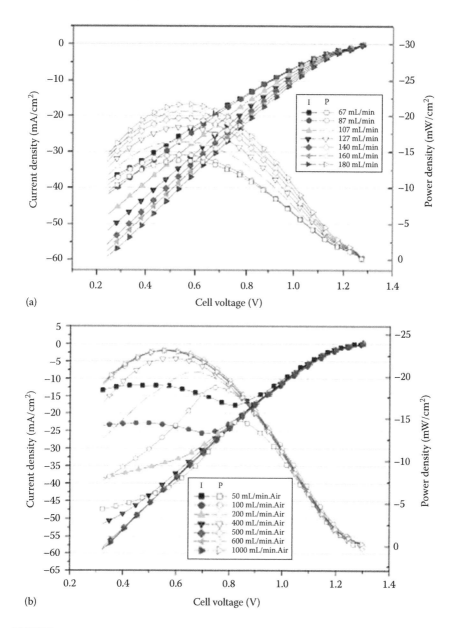

FIGURE 6.4 (a) Linear sweep voltammetry and power density curves at different (a) electrolyte 1.6 M V^{2+} (at the air flow rate of 500 mL/min) concentration (b) and (b) air flow rate (at the electrolyte flow rate of 73 mL V^{2+}/min); and (b) air flow rates (electrolyte concentration: 1.6 M V^{2+}; electrolyte flow rate: 73 mL/min V^{2+}). (Reprinted from *J. Power Sources*, 259, Noack, J., Cremers, C., Bayer, D., Tübke, J., and Pinkwart, K., A coupled-physics model for the vanadium oxygen fuel cell, 125–137. Copyright 2014, with permission from Elsevier.)

and electrolyte flow rates can significantly increase cell performance. For example, a maximum power density of 23 mW/cm^2 at a cell voltage of 0.56 V and a current density of 40 mA/cm^2. In addition, this cell could give a voltage efficiency of 33%, which is still insufficient when compared to other types of batteries. From the I–V polarization curves, it can be seen that the mass transfer limitation is not significant, but the ohmic drop should be responsible for a fast decrease in cell voltage with increasing current density. The voltage of the VOFC/2 M H$_2$SO$_4$ cell can be increased with increasing current density at the air flow rates of 100–1000 mL/min air (except 50 and 100 mL/min). However, current peaks are 18 mA/cm^2 and 26 mA/cm^2 at 0.82 and 0.7 V and air flow rates of 50 and 100 mL/min air, respectively, were obtained in VOFC/2 M H$_2$SO$_4$ due to a decrease in the ORR rate. It can also be observed that the current density decreases after the peak, probably due to the HER at low cathode potentials. The effect of air flow rate on VOFC performance diminishes when the air flow rate is faster than 200 mL/min. Voltammetric curves of VOFC at different flow rates of the V^{2+} electrolyte (67–180 mL/min) indicate that the activation and ohmic losses are the major factors affecting performance rather than transport limitations (with no limiting current densities) at all electrolyte flow rates. However, at the high electrolyte flow rates of 140, 160, and 180 mL/min, voltammetric curves are almost parallel, probably due to a polarization difference induced by the charge transfer. The maximum power density for VOFC/2 M H$_2$SO$_4$ was 18 mW/cm^2 at 0.22 V, 38 mA/cm^2, and the electrolyte flow rate of 67 mL/min. VOFC/2 M H$_2$SO$_4$ had a 43% higher activation loss and 33% higher resistance than those of VOFC/2 M EA [9].

Regarding the discharge capacity, this VOFC/2 M H$_2$SO$_4$ cell with a 280 cm^2 active area, shown in Figure 6.3, could yield a value of 49.5 Ah, representing 57.7% of the nominal (theoretical) discharge capacity due to 88% losses from the side HER, V^{2+} crossover into the intermediate space, and oxygen diffusion into a plastic container, ohmic (complex cell setup), and activation losses (water mass transport limitation on the three phase boundary of the cathode). However, it was observed that the rate of electrochemical oxidation from divalent to trivalent vanadium was higher than that of ORR at the cathode during discharge, evidenced by a lower anode overpotential. Electrochemical impedance spectroscopy was also used for studying the reaction kinetics in both charged and discharged states.

In addition, Austing et al. [10] developed a rechargeable VARFB (Figure 6.1b) with a bifunctional air-cathode (VARFB) that had a two-layer air-cathode. This electrode was similar to the BEs in zinc–air batteries, and it consisted of an oxygen reduction layer containing a GDL and Pt/C catalyst for discharging and graphite felt potion with IrO$_2$ as the catalyst for charging. This cell consisted of two half-cells separated by a proton exchange membrane. As claimed, this VARFB could combine the benefits of VRFBs and metal–air batteries. VARFB has a two-layered BE (one layer was hydrophobic for ORR and the other, hydrophilic layer for OER) at the positive electrode side, negative graphite felt electrode (for vanadium (II/III) oxidation/reduction during charging and discharging), current collector with flow field connected with BE, pumps for electrolyte circulation, and a tank with a vanadium electrolyte (anolyte) (V^{2+}/V^{3+}). Anolyte V^{3+}/2 M H$_2$SO$_4$ is purged with nitrogen to avoid the contact of the electrolyte with air. When discharging, the anolyte does not flow and contacts with the activated graphite felt, where V^{2+} was oxidized into V^{3+} (Reaction 6.1) and air passes through the flow field to the gas diffusion positive electrode, where ORR occurred (Reaction 6.2). When charging, the vanadium electrolyte (V^{2+}/V^{3+}) was circulated through the negative half-cell with the reduction of V^{3+} to V^2 and air was pumped through the flow field to the positive electrode, where OER occurred. Two electrodes were separated by an IEM, which enabled the diffusion of ions across the membrane while preventing crossover of the vanadium electrolyte. Obviously, this VARFB has the advantages of converting energy in the stack, storing energy in the tank, and using oxygen from air instead of the expensive vanadium electrolyte (43% of the total VRFB cost) in the positive half-cell. It was reported that this BVARFB with a two-layered cathode (Pt/C catalyst for ORR and IrO_2 catalyst for OER) could yield a power density of 34.6 mW/cm^2 at 40 mA/cm^2, with a maximum energy efficiency of 41.7% at the current density of 20 mA/cm^2 at room temperature [10]. Figure 6.5 shows the results of the cycle stability of such VARFBs.

However, it seemed that the VARFB described earlier should have a further improvement in blocking the crossover of vanadium ions. To further increase energy density, electrolyte additives may be needed for maintaining the maximal solubility of electroactive vanadium [10].

6.4 COMPONENTS OF VARFBs

6.4.1 ELECTRODES

The major function of the VARFB negative and positive electrodes is to provide active sites for electrochemical reactions, as expressed by Reactions 6.1 and 6.2, respectively. It is expected that both positive and negative electrodes should satisfy several requirements for high performance [1,12]. For the positive electrode, where ORR (during the discharging process) and OER (during the charging process) occur, high stability at high charging potential, low ORR overpotential during discharge, and high catalytic activity/stability for both ORR and OER are required. For the negative electrode, where the V^{3+}/V^{2+} redox reactions occur, high electrochemical activity/stability and the absence of side reactions, for example, hydrogen evolution (HER), are necessary. In addition, for both electrodes, some general requirements are (1) high electronic conductivity; (2) 3D structures having both a high contacting surface area and permeability toward the electrolyte; (3) high cycling stability; and (4) high chemical and mechanical stability. There are three main kinds of electrode materials: carbon-based materials, metal-based materials, and metalized graphite materials.

6.4.1.1 Negative Electrodes

VARFBs and VRFBs have similar electrochemical reactions at the negative electrodes. The negative electrodes of VARFBs are mainly made of carbon fiber [13], carbon felt [14–18], nitrogen-doped carbon [19,20], graphite plate [21], and graphite felt [12,22–24]. The carbon-based negative electrodes are typically stable due to their low electrode potentials. For example, using a graphite plate as the negative electrode could show favorable stability during several months of operation.

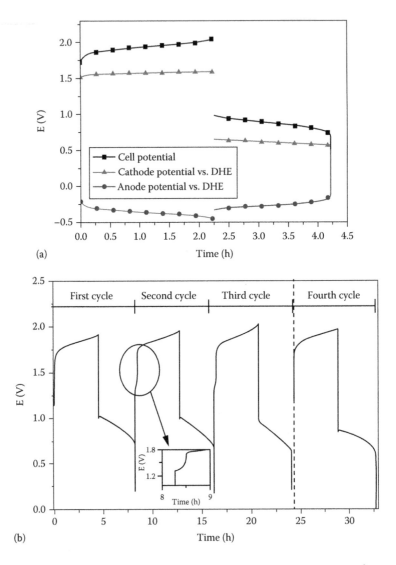

FIGURE 6.5   (a) Charge and discharge of bidirectional VARFB at 40 mA/cm². (b) Four cycles of VARFB at 20 mA/cm². Fourth cycle was recorder in fresh anolyte and catholyte. (Reprinted from *J. Power Sources*, 273, Ausiting, G., Kitcher, C.N., Hammer, E-M., Komsiyska, L., Wittstock, G., Hammer, M., Komsiyska, L., and Wittstock, G., Study of an unitized bidirectional vanadium/air redox flow battery comprising a two-layered cathode, 1163–1170. Copyright 2015, with permission from Elsevier.)

6.4.1.2   Positive Electrodes

Rechargeable VARFBs require a new type BE combining ORR and OER catalysts. Carbon/graphite-based felt could be used as the support. The electrochemical oxidation of carbon felt at high potentials of OER and ORR can especially result in chemical and mechanical disintegration, even after just a few cycles [24]. The felt produced from polymer PAN showed a higher resistance to oxidation than that of felt produced from rayon polymer [25]. It was believed that the rapid oxidation of rayon-based felt might be due to its microcrystalline structure [25]. The battery performance, a battery with carbon-based electrodes, a polystyrene sulfonic acid cation selective membrane, and 2 M H_2SO_4 containing 1.5 M sulfate vanadyl ($VOSO_4$) showed a coulombic efficiency of 90% and voltage efficiency of 81% at SOC [20]. The thermal activation of a felt electrode at 400°C (by covering its surface with certain functional groups, such as C=O and C–O) could increase its cycling life and energy efficiency [24]. It was believed that these C=O and C–O groups could be the active sites for vanadium redox reactions. Moreover, the chemical modification of graphite-based electrodes in sulfuric or nitric acids, or their mixture, could also produce oxygen functional groups on the graphite felt surface to facilitate the vanadium redox reactions. For example, a VRFB with graphite felt electrodes, treated in a boiling concentrated sulfuric acid for 5 h, could demonstrate a low resistance of 2.50 Ω cm^2 and energy efficiency of 91% [24]. A performance summary of main carbon-based electrodes is given in Table 6.2.

In the literature, metal-based electrode materials include Pt, platinized Ti, and IrO_2/TiO_2 [10,26–28]. Dimensionally stable anodes (DSAs), such as IrO_2/Ti, are relatively more stable in VARFBs [9,15,26]. An IrO_2-based DSA showed high stability over several charge/discharge cycles, with a charging current efficiency of over 90%. VRFBs with this DSA had 10%–90% of the SOC when the cell voltage was changed from 1.22 to 1.16 V, and they also had a stable OCV of 1.3 V for over 3 days at a temperature range of –5°C to 60°C. The most durable metal-based electrode for VRFBs is platinized electrode [26]. However, these metal-based electrodes are much more expensive than carbon-based electrodes. The rechargeable VARFB has a positive electrode in the form of the BE with an ORR and OER catalyst is the most advanced electrode. The ORR catalyst consists mainly of a carbon-supported catalyst (Pt/C), and the OER catalyst consists of IrO_2/Ti [10].

6.4.2 ION EXCHANGE MEMBRANES OF VARFBs

The membrane is one of the most expensive components of VARFBs, just as in VRFBs (Figure 6.2). The membranes of VARFB are also perfluorosulfonic acid polymers, for example, Nafion membranes (Dupont). They have both high ionic conductivity and good stability. However, due to its high cost (its price ranges from $500/m^2 to $800/m^2 [11]), many alternative cost-effective membranes, for example, SPEKK, were developed but with a compromised technical performance when compared with Nafion membranes. The primary functions of IEMs in VARFBs are to separate the negative electrolyte and facilitate the transport of protons. Requirements for VARFB's IEMs are as follows: (1) high selectivity and conductivity; (2) high electrochemical, chemical, and mechanical stability; and (3) well-managed water balance [29]. The main IEMs currently employed in VARFBs can be classified into two groups: perfluorosulfonated and nonfluoride IEMs.

One of the important issues in flow batteries is the crossover of both water and vanadium ions through the membrane due to its glassy properties and insufficiently interconnected proton conductive channels [30]. In particular, in diluted electrolytes,

TABLE 6.2
Performance Summary of VARFB Electrodes

Electrode	Vanadium Electrolyte	Battery Membrane	Battery Performance	References
Negative electrode				
VARFB five cells stack	3 M H_2SO_4	Nafion 115	Energy efficiency of 27.8%	[8]
Carbon fiber felt	2 M $VOSO_4$ in 4 M H_2SO_4	Anion–exchange	Energy efficiency of 80%	[10,13]
Carbon felt	2 M V(IV) in 2.5 M H_2SO_4 and 2.5 M V(III) in 2.5 M	Nafion/SiO_2	Energy efficiency of 80% at 20 mA/cm²	[10,29]
Carbon felt with chemical treatment	1.5 M $VOSO_4$ in 3 M H_2SO_4	Nafion	10 kW battery with energy efficiency of 80%	
Graphite felt	2 M V(III) in 2.5 M H_2SO_4	Nafion-PDDA-PSS	Energy efficiency 83.9% at charge and discharge currents of 80 and 20 mA/cm² respectively	[32]
Graphite plate	0.1 M V(III) in 2 M H_2SO_4	Sulfonated polyethylene anion membrane	Charge/discharge at 3 mA/cm²	[28]
Graphite-felt	2 M $VOSO_4$ in 2 H_2SO_4	Polysterene sulfonic acid cation-selective	22	[31]
Graphite felt bonded to graphite impregnated polyethylene plate	2 M $VOSO_4$ in 2 M H_2SO_4	Polysterene sulfonic acid	Energy efficiency of 87%	[31]
Graphite plate heat bonded to conducting plastic bipolar plate	1.5–2 M $VOSO_4$ in 2 M H_2SO_4	Selemion CMV	1 kW of 10 cells and electrode surface of 1500 cm²	[31]
Graphite felt after thermal treatment in air at 400°C for 30 h	3 M $VOSO_4$ in 2 M H_2SO_4	NA	Energy efficiency of 88%	[28]
Graphite felt with chemical treatment in boiling 5 M H_2SO_4	2 M V(III)/2 M H_2SO_4	NA	Energy efficiency of 91%	[15]
Positive electrode				
Bifunctional electrode (ORR—Pt/C, OER—IrO_2) (rechargeable battery) (rechargeable VARFB battery)	2 M V(III) in 2.5 M H_2SO_4	Nafion 112/115	28%	[10]

Source: Adapted from Skyllas-Kazacos, A. et al., *J. Electrochem. Soc.*, 158, R55, 2011. With permission from *Electrochemical Society.*

water crossover could cause the flooding of porous electrode [5]. In the literature, several commercial membranes were used in VARFBs, such as Nafion 115/212 [5,7–10] and crosslinked sulfonated poly(ether ether ketone) (SPEEK) [6].

As mentioned earlier, although Nafion-based membranes (DuPont, St Joseph, Missouri) are chemically stable and retain high proton conductivity in sulfuric acid electrolyte [1], their high cost can hinder their practical usage. Therefore, some efforts were made to replace such Nafion membranes with cost-effective and chemically stable nonfluorinated polyether ether ketone (PEEK)-based membranes for VARFBs [6]. Noask et al. [8] designed a VARFB with two membranes that had a cavity between them to minimize vanadium crossover. Their cell contained a membrane made of Nafion 115 in the anodic half-cell and a Nafion 212-based CCM in the cathodic half-cell. This VARFB demonstrates an average voltage efficiency of 33% and a discharge power density of 19.6 mW/cm^2.

Cost-effective nonfluorinated membranes, such as poly(ether ether ketone) (cSPEEK) membranes, were used in the modular VARFB with two different MEAs for charging and discharging, respectively. This VARFB demonstrated an energy efficiency of 26.67% at $T = 40°C$ and a current density of 2.4 mA/cm^2 for both charging and discharging [8]. Hosseiny et al. [5–7] improved the mechanical stability of SPEEK membranes by chemical crosslinking using electrophilic aryls as the crosslinker via the Friedel–Craft alkylation route. The crosslinking had two pathways: (1) crosslinking on the sulfonic acid groups and (2) crosslinking on the backbone. It was observed that the enhancement of crosslinking temperature increases the membranes' proton conductivity. High-temperature crosslinking could also result in an increase in proton conductivity and a 100 times lower VO^{2+} crossover than conventional Nafion membranes. It was observed that the water diffusion coefficient was increased in the membranes with increasing temperature. The vanadium permeability in cSPEEK membranes was dependent on the ratio of crosslinked and noncrosslinked SO$_3^-$ groups. The cSPEEK in VARFB that was tested had a proton conductivity of 27.9 mS/cm, which was a 100 times lower vanadium permeability than Nafion 117. Menitas and Skyllas-Kazacos [4] used a Nafion membrane in a five-cell VOFC and obtained the delamination of GDE from the membrane due to membrane swelling caused by water uptake.

6.4.3 VANADIUM ELECTROLYTE

The vanadium electrolyte is the second expensive components of VARFBs (the membrane being the first one) [9]. The impact of the price of V$_2$O$_5$ on VARFBs' energy cost is significant. For example, a 1 MW VRFB (8 MW h) system has an energy cost of between \$500/kW h and \$1200/kW h at the varying V$_2$O$_5$ price from \$5/kg to \$30/kg [31]. As discussed previously, VARFBs utilize V^{2+}/V^3 redox solution as an anolyte in the negative half-cell [32–47]. To increase the energy density of VARFBs, a high concentration of vanadium, usually 2–3 M in 2 M H$_2$SO$_4$, could be used as the anolyte. Furthermore, approaches for optimizing the anolyte V^{2+}/V^{3+} composition, thermal stability, and cost-efficiency should be necessary for VARFBs. VARFBs principally use a vanadium sulfate electrolyte (anolyte). For rechargeable VARFBs, a 1.2 M V^{3+}/V^{2+} anolyte was prepared electrolytically [10] in a traditional VRFB setup with a V^{2+}/V^{3+} anolyte and V^{4+}/V^{5+} catholyte [4,36]. Electrolysis was carried out in a VRFB setup with a solution containing 1.13 M $VOSO_4$, xH_2O, and 2.0 M H_2SO_4, and it was stopped at 1.8 V through two steps at 80 and 40 mA/cm^2, respectively, using twice more catholyte (by volume) than anolyte.

Normally, the solubility of V(IV), V(II), and V(III) compounds is small at low temperatures (the optimal range of $T = 10°C–40°C$) [40,41]. Vanadyl sulfate can reach a concentration of 4 M if a stabilizing agent or inhibitors of V (V) precipitation (such as hexametaphosphate, K_2SO_4, Li_2SO_4, and urea) are used [43,44]. As observed, increasing the concentration of H_2SO_4 to over 2 M can lead to VO precipitation in the V(III)/V(II) solution at $T < 10°C$ [40,41]. The thermal stability of the vanadium electrolyte could be improved with the addition of 5% K_2SO_4 into the electrolyte solution for 80 days at 4°C [43,44].

To increase vanadium concentration (4 M in 6 M H_2SO_4), thermal stability, cycle life, capacity, and efficiency, some gelled vanadium electrolytes could be developed using a gelling agent, such as fumed silica.

6.5   SUMMARY

VARFB technology is still at the lab prototype stage (technology readiness level [TRL 3–4]), mainly due to technical challenges, such as the rapid degradation of both expensive Nafion membranes caused by vanadium crossover and BEs with noble metal-based ORR and OER catalysts. However, this technology has significant potential for cost-effective energy storage due to a potentially doubled energy density caused by the elimination of the catholyte tank, in addition to the free cathode reactant (air).

REFERENCES

1. Parasuramn, A., Lim, T.M., Menictas, C., and Skyllos-Kazacos, M. 2013. Review of material research and development for vanadium redox flow battery applications. *Electrochimica Acta*, 101: 27–40.
2. Fabjan, Ch., Garche, J., Harrer, B., Jörissen, L., Kolbeck, C., Philippi, F., Tomazic, G., and Wagner, F. 2011. The vanadium redox-battery: An efficient storage unit for photovoltaic systems. *Electrochimica Acta*, 47: 825–831.
3. Kaneko, H., Negishi, A., Nozaki, K., Sato, K., and Nakaimi, M. 1994. Redox battery, U.S. Patent 5318865.
4. Menitas, C. and Skyllas-Kazacos, M. 2011. Performance of vanadium-oxygen redox fuel cell. *Journal of Applied Electrochemistry*, 41: 1223–1232.
5. Hosseiny, S.S., Saakes, M., and Wessling, M. 2011. A polyelectrolyte membrane-based vanadium/air redox flow battery. *Electrochemical Communications*, 13: 751–754.
6. Hosseiny, S.S., Ioanan, F.C., Demco, D.U., Saakes, M., and Weaseling, M. 2014. *Membranes*, 24: 41–59.
7. Hosseiny, S.S. 2008. Doctor Philosophy Thesis, University of Trent, Peterborough, Ontario, Canada.
8. Noack, J., Cremes, C., Bayer, D., Tubke, J., and Tuebke, J. 2010. Air breathing vanadium/oxygen fuel cell, in *Proceedings of the 218th ESC Meeting Abstracts*, 218th ECS meeting abstracts 2010, Vol. 1. Las Vegas, NV, October 10–15, 2010.

9. Noack, J., Cremes, C., Bayer, D., Tubke, J., and Pinkwart, K. 2014. A coupled-physics model for the vanadium oxygen fuel cell. *Journal of Power Sources*, 259: 125–137.

10 Austing, J.G., Kitcher, C.N., Hammer, E.-M., Komsiyska, L., and Wittstock, G. 2015. Study of an unitized bidirectional vanadium/air redox flow battery comprising a two-layered cathode. *Journal of Power Sources*, 273: 1163–1170.

11. Viswanathan, V., Crawford, A., Stephenson, D., Kim, S., Wang, W., Li, B., Coffey, G. et al. 2014. Cost and performance model for redox flow batteries. *Journal of Power Sources*, 24: 1040–1051.

12. Xi, J., Wu, Z., Qiu, X., and Chen, L. 2007. Nafion/SiO$_2$ hybrid membrane for vanadium redox flow battery. *Journal of Power Sources*, 166: 531–536.

13. Kageyama, Y., Tayam, T., and Sato, K. 1997. U.S. Patent 5656 1997.

14. Chieng, S.C., Kazacos, M., and Skyllas-Kazacos, M. 1992. Preparation and evaluation of composite membrane for vanadium redox battery applications. *Journal of Power Sources*, 39: 11–19.

15. You, D., Zhang, H., and Chen, H. 2009. A simple model for the vanadium redox battery. *Electrochimica Acta*, 54: 6827–6836.

16. Teng, X., Zhao, Y., Xi, J., Wu, Z., Qiu, X., and Chen, L. 2009. Nafion/organically modified silicate hybrids membrane for vanadium redox flow battery. *Journal of Power Sources*, 189: 1240–1246.

17. Jia, C., Liu, J., and Yan, C. 2010. Effect of electro-oxidation current density on performance of graphite felt electrode for vanadium redox flow battery. *Journal of Power Sources*, 195: 4380–4383.

18. Chieng, S.C. and Skyllas-Kazacos, M. 1992. Modification of daramic, microporous separator, for redox flow battery applications. *Journal of Membrane Science*, 75: 81–91.

19. Shao, Y.Y., Wang, X.Q., Engelhard, M., Wang, C.M., Dai, S., Liu, J., Yang, Z.G., and Lin, Y.H. 2010. Nitrogen-doped mesoporous carbon for energy storage in vanadium redox flow batteries. *Journal of Power Sources*, 195: 4375–4379.

20. Sangki, P. and Hansung, K. 2015. Fabrication of nitrogen-doped graphite felts as positive electrodes using polypyrrole as a coating agent in vanadium redox flow batteries. *Journal of Materials Chemistry*, 3: 12276–12283.

21. Skyllas-Kazacos, M., Rychci, M., Robins, R., Fane, A., and Green, M. 1996. Vanadium redox cell electrolyte optimization studies. *Journal of the Electrochemical Society*, 133: 1057–1058.

22. Xi, J., Wu, Z., Teng, X., Zhao, Y., Chen, L., and Qiu, X. 2008. Self-assembled polyelectrolyte multilayer modified Nafion membrane with suppressed vanadium ion crossover for vanadium redox flow batteries. *Journal of Materials Chemistry*, 18: 1232–1238.

23. Qian, P., Zhang, H., Chen, J., Wen, Y., Luo, Q., Liu, Z., You, D., and Yi, B. 2008. A novel electrode-bipolar plate assembly for vanadium redox flow battery applications. *Journal of Power Sources*, 175: 613–619.

24. Chakrabarti, M., Roberts, E., and Saleem, M. 2010. Charge–discharge performance of a novel undivided redox flow battery for renewable energy storage. *International Journal Green Energy*, 7: 445–460.

25. Zhong, S., Padeste, C., Kazacos, M., and Skyllas-Kazacos, M. 1993. Physical chemical and electrochemical properties comparison for rayon and PAN based graphite felt electrodes. *Journal of Power Sources*, 45: 29–41.

26. Zhong, S. and Skyllas-Kazacos, M. 1992. Electrochemical behavior of vanadium(V)/vanadium(IV) redox couple at graphite electrodes. *Journal of Power Sources*, 39: 1–9.

27. Rychcik, M. and Skyllas-Kazacos, M. 1987. Evaluation of electrode materials for vanadium redox cell. *Journal of Power Sources*, 19: 45–54.

28. Sun, B.T. and Skyllas-Kazacos, M. 1991. Chemical modification and electrochemical behaviour of graphite fibre in acidic vanadium solution'. *Electrochimica Acta*, 36: 513–517.

29. Skyllas-Kazacos, M., Chakrabarti, M.H., Hajimolana, S.A., Mjalli, F.S., and Saleem, M. 2011. Progress in flow battery research and development. *Journal of the Electrochemical Society*, 158: R55–R79.

30. Hosseiny, S.S. and Wessling, M. 2011. Ion exchange membranes for vanadium redox flow batteries, in *Advanced Membrane Science and Technology for Sustainable Energy and Environmental Applications*, A. Basile and S. Nunes (eds.). Woodhead Publishing, Cambridge, U.K., pp. 413–434.

31. Aoron, A., Liu, Q., Tang, Z., Grim, G.M., Papandrew, A.B., Turhan, A., Zawodzinski, T.A., and Mench, M.M. 2012. Dramatic performance gains in vanadium redox flow batteries through modified cell architecture. *Journal of Power Sources*, 206: 450–453.

32. Sukkar, T. and Skyllas-Kazacos, M. 2003. Modification of membranes using polyelectrolytes to improve water transfer properties in the vanadium redox battery. *Journal of Membrane Science*, 222: 249–264.

33. Zeng, J., Jiang, C.P., Wang, Y.H., Chen, J.W., Zhu, S.F., Zhao, B.J., and Wang, R. 2008. Membranes for redox flow battery applications. *Electrochemistry Communications*, 10: 372–375.

34. Teng, X., Zhao, Y., Xi, J., Wu, Z., Qiu, X., and Chen, L. 2009. Nafion/organic silica modified TiO_2 composite membrane for vanadium redox flow battery via in situ sol-gel reaction. *Journal of Membrane Science* 234: 149–154.

35. Sang, S., Wu, Q., and Huang, K. 2007. Preparation of zirconium phosphate (ZrP)/Nafion1135 composite membrane and H^+/VO^{2+} transfer property investigation. *Journal of Membrane Science*, 305: 118–124.

36. Chen, D., Hickner, M.A., Agar, E., and Kumbur, E.C. 2013. Optimizing membrane thickness for vanadium redox flow batteries. *Journal of Membrane Science* 437: 108–113.

37. Luo, X., Lu, Z., Xi, J., Wu, Z., Zhu, W., Chen, L., and Qiu, X. 2005. Influences of péremption of vanadium ions through PVDF-g-PSSA membranes on performances of vanadium redox flow batteries. *Journal of Physical Chemistry*, B 109: 20310–20314.

38. Mohammadi, T. and Kazacos, M.S. 1996. *Journal of Power Sources*, 63: 179–186.

39. Jian, X.G., Yan, C., Zhang, H., Zhang, S.H., Liu, C., and Zhao, P. 2007. Synthesis and characterization of quaternized poly(phthalazinone ether sulfone ketone) for anion-exchange membrane. *Chinese Chemical Letters*, 18: 1269–1272.

40. Kazacos, M., Cheng, M., and Skyllas-Kazacos, M. 1990. Electrolyte optimization of vanadium redox cell. *Journal of Applied Electrochemistry*, 20: 463–467.

41. Skyllas-Kazacos, M., Menictas, C., and Kazacos, M. 1996. Thermal stability of concentrated V(V) electrolytes in the vanadium redox cell. *Journal of the Electrochemical Society*, 143: L86–L88.

42. Rahman, F. and Skyllas-Kazacos, M. 2009. Vanadium redox battery: Positive half-cell electrolyte studies. *Journal of Power Sources*, 189: 1212–1219.

43. Skyllas-Kazacos, M., Peng, C., and Cheng, M. 1999. Evaluation of precipitation inhibitors for supersaturated vanadyl electrolytes for the vanadium redox battery. *Electrochemical and Solid-State Letters*, 2: 121–124.

44. Samad, M.A. and Skyllas-Kazacos, M. 1994. Recent progress with the UNSW vanadium battery, in *Proceedings of the Ninth Australasian Electrochemical Conference*, Wollongong, New South Wales, Australia, February 1994, pp. 033-1–033-4.

45. Rahman, F. and Skyllas-Kazacos, M. 1998. Solubility of vanadyl sulfate in concentrated sulfuric acid solutions. *Journal of Power Sources*, 72: 105–110.

46. Oriji, G., Katayama, Y., and Miura, T. 2004. Investigation on V(IV)/V(V) species in a vanadium redox flow battery. *Electrochimica Acta*, 49: 3091–3096.

Index

Printed and bound by CPI Group (UK) Ltd, Croydon, CR0 4YY

01/11/2024

01782619-0019